Apparel Machinery and Equipments

Apparel Machinery and Equipments

R.Rathinamoorthy and R. Surjit

WOODHEAD PUBLISHING INDIA PVT LTD

New Delhi, India

Published by Woodhead Publishing India Pvt. Ltd.
Woodhead Publishing India Pvt. Ltd., 303, Vardaan House, 7/28, Ansari Road,
Daryaganj, New Delhi - 110002, India
www.woodheadpublishingindia.com

First published 2015, Woodhead Publishing India Pvt. Ltd.
© Woodhead Publishing India Pvt. Ltd., 2015
Reprinted 2020

Woodhead Publishing India Pvt. Ltd. ISBN: 978-9-38030-859-3
Woodhead Publishing Ltd. e-ISBN: 978-9-38030-815-9

Typeset by Third EyeQ Technologies Pvt Ltd, New Delhi
Printed and bound by Replika Press Pvt. Ltd.

Contents

Acknowldgement

We authors take this opportunity to acknowledge the people who have helped us writing the book. First of all we would like to thank the almighty for giving us strength and knowledge to write a book. We would like to thank Mr. GANDHIRAJAN, Manager sales, M/s. Mehala Machines, Tirupur and Mr. BASANTA KUMAR MOHANTY, Country Manager – India, Operations Manager (Sales), South Asia, M/s. Macpi Group (Banglore) for their valuable inputs and support in writing this book and permitting us to use their machinery photographs for explanation purpose. We would like to place on record a special thanks to our Principal and the PSG College of Technology management for providing us the atmosphere and guidance to come out with such a work. We thank Mr.R.P.SUNDARAM, Associate Professor, Department of Textile Technology, PSG College of Technology, for writing the foreword. Last but not least, we would like to acknowledge the support rendered by our family members for writing this book.

Foreword

It gives me immense pleasure to pen few lines about the book and the authors. This book on Apparel machinery and equipment looks brilliant in its first sight. I would like to congratulate the authors for coming out with such a book as it is the need of the hour for apparel technologists to have in depth knowledge on the machines used in apparel. This book gives adequate knowledge on cutting machines, sewing machines, sewing attachments, feeding mechanisms, lubrication systems, and maintenance of machines and also finishing machines are described at length.

Fashion designing is a brilliant expression of once own individuality and is largely reserved for the celebrities, the elites of the society. This is largely due to the knowledge gap and the desirability gap. The knowledge gap is the gap between what is possible by the 'state of the art technologies and technical equipment' of the Apparel industry and what is actually realized. This knowledge gap can only be bridged by comprehensive treatise like this one. Though the technical manuals of the machinery manufactures are helpful in this regard, they are too detailed and lack 'comprehensive knowledge in a nut shell'. Thus the well informed manufacturing personals will usher the merger of the 'scalability success factor' of the garment industry with the 'psychological and emotional success factors' of the apparel industry. The desirability gap will get reduced as the common customer tastes the luxury of the fashions as it becomes affordable. By this the ultimate success factor of 'customer demand' will make the apparel industry as a real ubiquitous industry.

The authors Mr. Rathinamoorthy and Mr.Surjit have been meticulous in their approach of writing this book with appropriate content. Mr. R. Rathinamoorthy was my post graduate student and to write a foreword for his book makes me feel elated. I am sure he is going to write many more high quality books like these which will benefit the stakeholders and academia. Mr. R.Surjit has been known to me since five years and this book shows his potential and the knowledge related to apparel industry. His industrial experience has come in handy and it becomes evident as I read this book.

The book is well written with more emphasis on sewing machines. The book contains more illustrations to give better understanding for the readers. The best part of this book is its detailed coverage on various setting points in sewing machines. It is first of its kind in this area with so much depth in the sewing machinery. Clear illustrations are provided for various setting points which can be used by industrial personnel also. It will surely impart practical knowledge to the readers. I am sure this book will be handy for the students, academicians and industry personnel. I wish good luck for the authors and look forward for many more similar comprehensive works.

R.P.Sundaram

Spreading and cutting

This chapter explains the basic criteria that needs to be considered while spreading the fabric for cutting process and explains the different machine used in the spreading process. The chapter also explains the common cutting machineries used in the apparel industry with their commercial importance. The latest and advanced cutting machines are also explained with their merits and demerits.

Keywords: Spreading, Requirements, Straight knife, Advanced cutting machine, Applications.

1.1 Introduction to apparel industry

The apparel industry is one of the oldest and largest industry providing ample employment opportunities and it exemplifies the growth in global manufacturing. This industry is very versatile in nature and offers the world with a choice of garments ranging from mass market to high end fashion. This industry follows a combination of functional and line type of organization. The various departments of the apparel industry are shown in Figure 1.1.

Figure 1.1   Departments of apparel industry.

The merchandising department takes care of customer relations and receives order from the buyer followed by sampling department which prepares the

samples for the buyer. There are various types of samples prepared as per
the requirement of the buyer and once the order is received, the fabric store
department and trims, accessories store department work on getting the
required material for starting the production. This process is followed by
spreading, cutting, sewing and washing (if required). Once the garment is
sewed, quality assurance department comes into work and checks the garment
quality and sees whether it matches the requirements of the customer. Then
the finishing and packing departments pack and make the goods ready for
shipment. Along with these departments, allied departments like maintenance,
HR, accounts, finance and purchase aid in the successful working of a garment
industry. In Figure 1.2, the process flow of an apparel industry is given for
better understanding. Buyer–supplier meeting leads to production order being
placed. Sampling is then done and then bill of materials (BOM) is generated
for raw material procurement and once raw material is ready, its inspection is
carried out. Washing is done followed by preparing pre-production sample.
On approval, the production is done and goods are kept ready for inspection
followed by finishing, packing and dispatch.

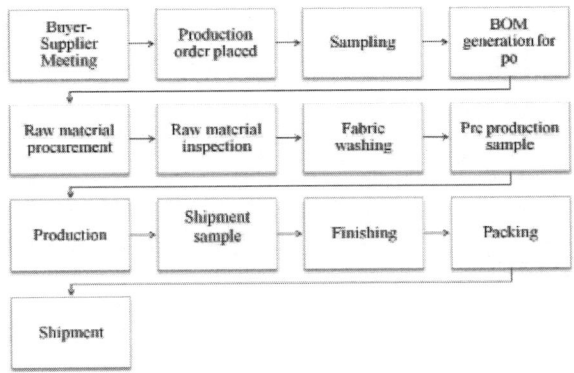

Figure 1.2   Process flow of an apparel industry.

This industry involves lot of machinery working as machines form an
integral part. The machines range from spreading and cutting to various
folding and finishing machines. Sewing machinery range from single needle
lockstitch machine to various types of special purpose machinery and they
are very essential for making versatile garments. It is a labour intensive and
machinery intensive industry and we are going to understand the various
machinery used in this industry in detail.

1.2 Spreading

This is a preparatory operation for cutting and consists of laying plies of one
cloth on top of the other in a predetermined direction and relationship between

the right and wrong sides of the cloth. The composition of each spread, i.e. the number of plies of each colour is obtained from the cut order plan.

Number of plies depends on:

1. Capacity of the cutting machine,
2. Volume of production,
3. Type of fabric itself (rough or slippery) and
4. Thickness of fabric.

Spreading fabric for cutting may be done in a variety of ways. These spreading modes describe the way in which the face of the fabric will be oriented, and what the nap direction is from ply to ply. The choice of spreading mode will affect the cost of spreading and the quality of the finished product (the result of the cutting).

Spreading quality is achieved when any flaws in the face of the fabric can be identified by the spreader (even if the fabric was pre-inspected), and removed (either during the process of spreading or marked for removal after spreading). The highest levels of spreading quality are, therefore, achieved with spreading modes that permit the face of the fabric to be 'up' and visible to the spreader at all times.

1.2.1 Spreading methods

The spreads can be of two basic types:

1. Flat spreads—all plies are of the same length.
2. Stepped spreads—this as the name suggests, is built up in steps, with all the plies in one step having the same length. A stepped spread is generally used when the quantities to be cut precludes the use of a flat spread. The cut order plan details the colours and ply lengths for a stepped spread, if it is needed.

1.2.2 Requirements of fabric spreading

1. Alignment of fabric plies: Every ply should comprise at least width of the marker plan, but should have the minimum possible extra outside those measurements. The textile materials vary in width. The marker plan is made of fit the narrowest width. In accuracy in this alignment, it could mean that plies do not cover the whole area of the marker plan and parts of some pattern pieces would be missing when cut.

2. Correct ply tension: The ply tension should be correct. If the tension is low then there will be ridges in the plies and if the tension is too

high then the fabric may shrink after cutting and sewing. The use of spreading machine gives uniform tension.

3. Fabric must be flat: The fabric laid on the table should be flat otherwise there will be ridges in it.

4. Elimination of fabric faults: Fabric faults (holes, stains, etc.) may be identified by the fabric supplier and additional faults may be detected during examination of fabric by the garment manufacturer prior to spreading. The spreader cuts across the ply at the position of the fault and pulls back the cut end to overlap as far back on the next splice mark. Splice marks are marked on the edge of the spreading table prior to spreading, by reference to the marker and ensure that whenever a splice is created the overlap of fabric is sufficient to allow complete garment parts rather than sections only to be cut. Computerized methods of achieving this are now available which provide a display of the marker plan on a computer screen on the spreading machine.

5. Correct ply direction and adequate lay stability: These two factors must be considered together. They depend on fabric type, pattern shape and the spreading equipment that is available. When the pattern pieces have been positioned in a particular direction in the marker plan, it is essential that the fabric is spread in a way that maintains that direction. It could be because of the effects of a surface design or fabric construction, problems of instability with a nap or a pile surface. This can require that some fabrics are spread with all the plies face. Symmetrical pattern pieces are placed all the same way up or face to face. If the pattern pieces are asymmetrical all the pieces face up or face down.

6. Elimination of static electricity: In spreading plies of fabric containing man-made fibres, friction may increase the charge of static electricity in the fabric. Friction may be reduced by changing the method of threading the fabric through the guide bars of the spreading machine. Humidity in the atmosphere of the cutting room may also be increased, thus allowing the static electric to discharge continuously through the atmosphere. In some case it may be necessary to earth the lay.

7. Avoidance of fusion of plies: In case of thermoplastic fibre fabrics, they may fuse together during cutting if the cutting knife becomes hot. We can prevent fabric from fusion by:

    (a) Using anti-fusion paper,

    (b) Using silicon lubricants on the knife blade,

    (c) Reducing ply height.

8. Avoidance of distortion in spread: There should not be friction between the bottom of the spread and the surface of the table. So a layer of hard polyethylene sheet is laid at the bottom of the spread.

9. Easy separation of the cut lay into bundles: Identification marks are used into plies due to colour or shade variation of fabric or other cases. For this separation, low valued coloured paper is used to plies.

10. Matching checks or stripes: If the fabric is checked or striped then it must be laid to the marker plan and they should be matched by the help of needle.

1.2.3 Modes of spreading

1. *Face/One/Way, Nap/One/Way mode of spreading (F/O/W, N/O/W)*

The highest quality of spreading is achieved by the Face/One/Way, Nap/One/Way mode of spreading (F/O/W, N/O/W) as shown in Figure 1.3. Each layer of fabric is spread with the face up (usually) permitting the spreader to see all of the face of the cloth to identify any flaws in the fabric. The fabric is spread in one direction only, from the end of the table to the beginning (Right to Left from the machine operators point of view). This will ensure that there will be no problems with nap direction in the finished product. For this mode of spreading, the patterns in an open marker are placed N/O/W. This is a slow method of spreading however, because after each layer is spread, the fabric is cut at the end (across the width of the table just past the beginning of the marker), and the machine and operator transverse back to the opposite end of the table to begin spreading the next layer of fabric (known as 'deadheading', a term borrowed from the trucking industry to mean 'travelling without a load'). This process is repeated until all the plies needed are spread.

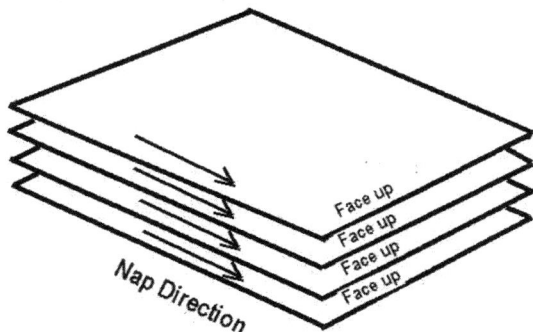

Figure 1.3   Face/ One/ Way, Nap/ One/ Way mode of spreading.

2. *Face/One/Way, Nap/Up/and Down method spreading (F/O/WN/U/D)*

The second highest level of spreading quality is possible with the Face/ One/Way, Nap/Up/and Down method of spreading (F/O/WN/U/D). In this mode, the fabric is spread from the end of the table to the beginning. At the beginning of the table, the spreader cuts the fabric across the width and then it must rotate the roll of fabric 180° (in the same plane). The spreader then continues spreading the fabric from the beginning back to the end of the table where the fabric will be cut and rotated again. This process is repeated until all the plies needed are spread. This mode requires that the fabric be symmetric, as alternating plies are placed in opposite directions. Markers for this method are most often open, Nap/Up/ Down to take advantage of asymmetric fabric, and are more efficient (than Nap/One/Way) (Figure 1.4). The marker may be Nap/One/Way although there will be no gain in quality (the only gain would be more efficient spreading time).

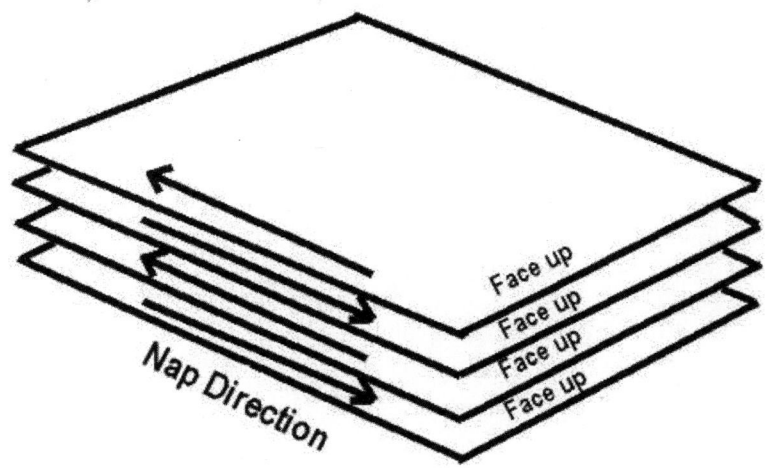

Figure 1.4   Face/ One/ Way, Nap/ Up/ and Down method spreading.

3. *Face to Face, Nap/Up/Down (F/F-N/U/D)*

The most efficient (fastest) method of spreading is the second lowest quality method. Face to Face, Nap/Up/Down (F/F-N/U/D). For symmetric fabrics and moderate overall quality, this method of spreading is widely popular. Starting at the end of the table, the spreader spreads the fabric to the beginning of the table. Without cutting the end, the spreader folds over and weights the fabric end down, and begins spreading back towards the end again (see Figure 1.5)

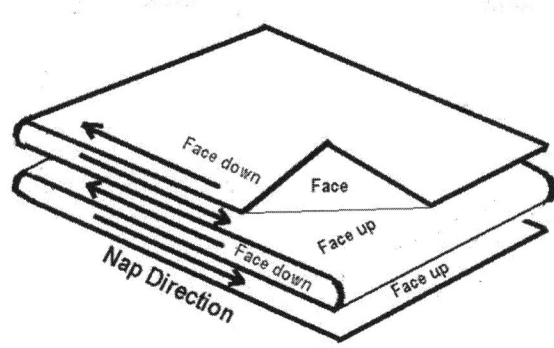

Figure 1.5   Face to Face, Nap/ Up/ Down (F/F- N/U/D).

For open, Nap/Either/Way markers, this mode produces the lowest cost of spreading and most efficient (least costly) fabric consumption. The quality is low, as the face of every other ply is not visible to the spreader to see and remove damages. This mode of spreading also facilitates the use of closed markers on open fabric, Nap/Either/Way, Nap/Up/Down or Nap/One/ Way (most moderate and least efficient, respectively). This mode requires the identification of damage parts during the sewing process by the sewing operators or quality.

4.   *Face to Face, Nap/One/Way (F/F-N/O/W)*

When fabric is asymmetric, the Face to Face, Nap/One/Way mode allows the use of an open or closed marker on open fabric that is napped or one-directional (Figure 1.6). The result of this mode is fabric that is face to face, where consecutive plies will yield pairs of parts (left and right). This method is relatively slow, and produces the lowest quality, as the face of every other ply is not visible to the spreader. This mode also requires the identification of damage parts during the sewing process by the sewing operators or quality control inspectors.

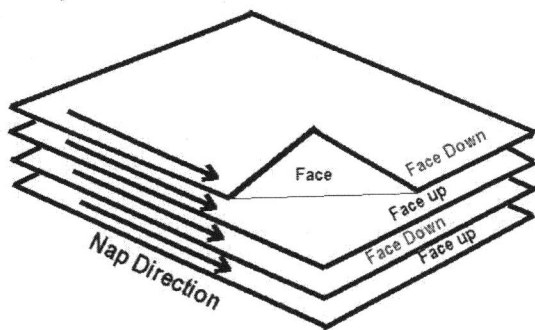

Figure 1.6   Face to Face, Nap/ One/ Way (F/F- N/O/W).

5. *Closed-Face to Face, Nap/One/Way*

Closed fabric is fabric that is folded in its length either due to the textile milling process (as with tubular knits), or deliberately by the mill to facilitate the manufacturing process. (Very wide fabrics might be purchased folded to enable the use of narrower tables for spreading that are already in place in the facility). Closed-Face to Face, Nap/One/Way spreading is the process where the spreader starts at the end of the table spreading the fabric (tubular or folded and rolled) back to the beginning of the table. The spreader cuts across the fabric width past the marker end, and then transverses back to the end of the table to start the process again. Two layers of fabric are laid on the table in one pass, where both layers are Face/Face (Figure 1.7).

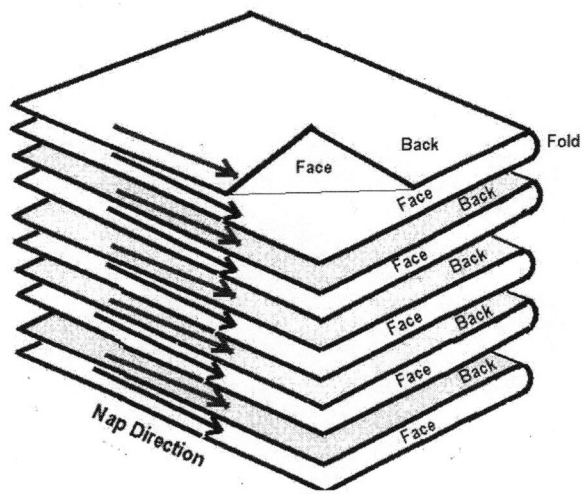

Figure 1.7   Closed – Face to Face, Nap/ One/ Way.

Folded fabrics facilitate the use of closed markers where the fold is utilized for parts that are single, in conjunction with paired parts. (A single back panel pattern is folded in half, and is laid on the edge of the fold. Other, paired parts are placed in the open areas of the marker, and when cut, yield left and right pairs.) Use of the closed marker (half a set of patterns) speeds the cutting process, as it takes roughly half the time to cut half a set of patterns. Quality, as other Face/Face modes is moderate at best, as half the fabric spread is not viewable by the spreader.

6. *The Closed-Face/Face, Nap/Up/and Down*

The Closed-Face/Face, Nap/Up/and Down mode is similar to the Closed-Face to Face, Nap/One/Way mode except that after the first pass, the spreader does not cut the fabric off at the beginning of the table. Instead, the fabric (two plies Face to Face) is folded over and the spreader and begins spreading back to the end of the table (Figure 1.8). This results in a Face to Face mode where *pairs of plies alternate up and down the table*. As this method reverses the direction of the nap, the quality is lower. Unless the nature of the nap or construction is such that in the end use of the product, the nap direction is not noticeable by the consumer, this method would not be used.

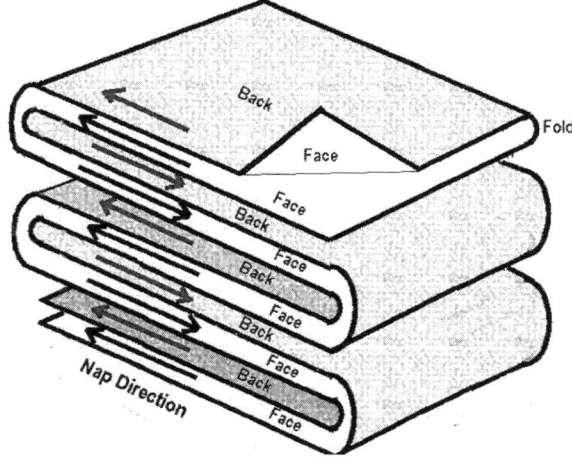

Figure 1.8   The Closed – Face / Face, Nap/ Up/ and Down.

1.2.4 Spreading methods

1.2.4.1 Manual spreading

1. **By hand:** Fabric roll is spread on the table by two workers according to the length and width of marker as shown in Figure 1.9. Sometimes it is done by entering a rod which is made by wood or metal in the centre of paper table which is present in the centre of the fabric roll.

Figure 1.9   Manual Spreading process.

2. **By hook:** In this process the top of the table on which the fabric is spread is set at 10° angle perpendicularly. The hook which is placed on the upper face of the table is 15 cm long and the hook is joined with one end of selvedge of the fabric. After completing the spreading of fabric, the top of the table is set again. The hooks are displaced and the marker is spread on the fabric lay.

3. **Spreading truck with the help of operator:** There is a spreading truck on the one end of the spreading table in which the fabric roll is placed. Then the truck is operated by hand from one end to the other end of the table and with the same time the fabric is open out from the fabric roll and the fabric is spread according to the length and width. This process is given in Figure 1.10.

Figure 1.10   Manual Spreading process with spreading truck.

Advantages

*   Can be used for low quantity, the capital cost is less

*   While spreading strips and checked, the lines can be matched

Disadvantages

*   Low productivity

*   Speed of the spreading depends up on the operator skill

*   The ply tension control varies at different place since more than one people do the spreading at the same time

1.2.4.2 Computerized spreading machine (automated)

Figure 1.11 shows a computer controlled spreading machine with automatic spreading mechanism. The working of the machine is detailed below.

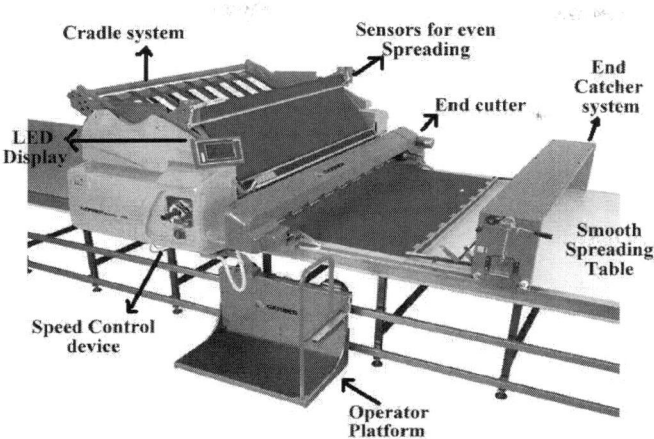

Figure 1.11   Computer controlled spreading machine.

Working

- Spreading machines carry the piece of fabric from end to end of the spread, dispensing one ply at a time onto the spread. Their basic elements consist of a frame or carriage, wheels travelling in guide rails at the edge of the table, a fabric support and guide collars to aid the correct unrolling of the fabric.

- In the simpler versions, the operator clamps the free end of fabric in line with the end of the spread, pushes the spreader to the other end, cuts off the ply in line with that end, clamps the beginning of the next ply, pushes the spreader to the other end and so on.

- More advanced spreading machines may include a motor to drive the carriage, a platform on which the operator rides, a ply-cutting device with automatic catcher to hold the ends of the ply in place, a ply counter, an alignment shifter actuated by photo-electric edge guides, a turntable and a direct drive on the fabric support, synchronized with the speed of travel, to reduce or eliminate tension in the fabric being spread.

  - The maximum fabric width that can be handled is normally 2 m. Extra wide machines capable of handling up to 3 m are available.

  - The maximum weight of cloth roll that can be carried by the larger spreading machines is 120 kg.

  - The maximum spreading speed around 100 m/min and the maximum height of spread cloth 28 cm.

  - When a spreading machine dispenses fabric when travelling in one direction but returns to the first end without spreading to begin the next ply, the return pass is known as 'dead heading'. Many spreaders will travel at a considerably higher speed when 'dead heading'.

- The advent of microprocessor control has enabled the development of more automatic functions on spreading machines. Thus a spreader can be preset to a selected number of plies, emitting an audible signal when it has reached the selected number or has come to the end of a piece of fabric.

Parts and functions

1. **Cradle system**

   - The cradle is the feed system for the spreading machine, where the fabric roll can be loaded for the spreading purpose (Figure 1.12).

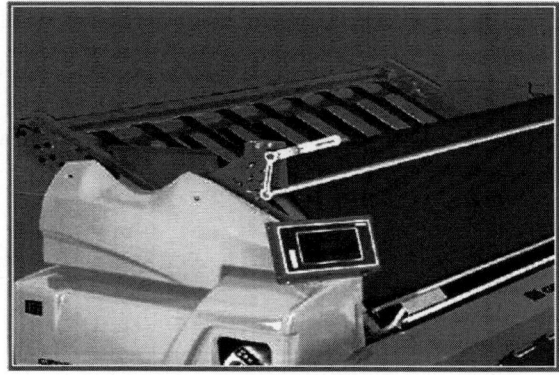

Figure 1.12   Cradle system in spreading machine.

   - These systems are electrical or mechanical, both the cases, cradles are tiltable for easy loading of heavy weight cloth rolls in it

   - Cradles are more effective and are particularly useful for materials which are hard to unwind or have a tendency to stretch or to come off the roll in an uneven manner.

2. **Sensors for even spreading**

   - Tensioning involves synchronizing the rate of spreading with the rate fabric is unrolled. A positive feed system utilizes a covered roller that is driven and timed to the movement of the machine. It prevents the momentum of a large roll from continuing to unwind when the machine slows down or stops. Roller covers of different materials may be used to give better gripping power for different types and weights of fabric.

   - Positioning devices and sensors monitor position and control fabric placement during spreading. These devices improve the quality of a spread. Electronic edge sensors monitor selvedges as fabric is spread. A deviation from the proposed alignment triggers a motor that shifts the roll to the correct position. Alignment can be held to 1/8 in. tolerance with these devices (Figure 1.13).

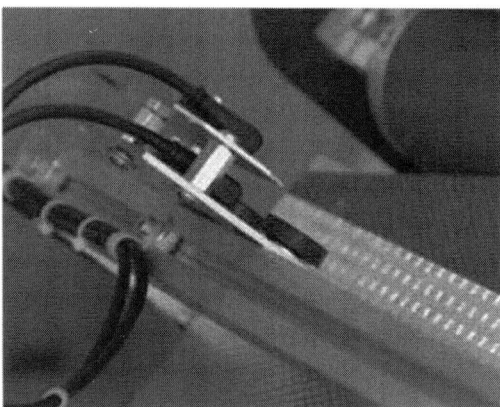

Figure 1.13   Sensors for the proper ply alignment and even ply tension.

- Width indicators may sound an alarm to alert the operator whenever fabric becomes narrower than the established width. Width variations are analyzed to determine where in the marker they fall, whether the fabric will still fit the marker, or whether the variation should be treated as a defect and removed.

3. **Damage control mark sensors**

One of the newest advances in spreading systems is the use of automatic sensors and marks on the piece goods to identify damages in the fabric. Reflective tape is applied to the fabric selvedge during the pre-inspection process. Automatic spreading machines are mounted with a sensor that detects the tape as it passes through the electric edge control eye. The sensor stops the spreading machine permitting the spreader to locate the damage and remove it during the spreading process. This technology is particularly useful for Face/Face modes of spreading where the spreader cannot see the face of the fabric on every other ply. This system can essentially assure the same spreading quality from F/F as F/O/W spreading. Also, this system permits higher spreading speeds as the spreader is not limited to how fast they can spread and see damages at the same time.

4. **Fabric cutter**

- A cutting knife (mostly round knife) is used to attach in front of the moving spreader and it cuts the fabric once the predetermined length is spread by moving across the table width.

- The knife may cut in one direction or both based on the type of material used. The return dead head of the knife speed will be maximum when compared to cutting speed (up to 200 m/min).

- The knife cutting length can be adjusted in control panel based on the fabric width cut.

- An automatic height detection sensor ensures the minimum distance between the cutting device and the table top, so avoiding wrinkles in the laying process.

5. **End catcher**

End catcher is a special set up either permanently fixed in the spreading table, or movable or fixed with the spreader head as in Figure 1.14.

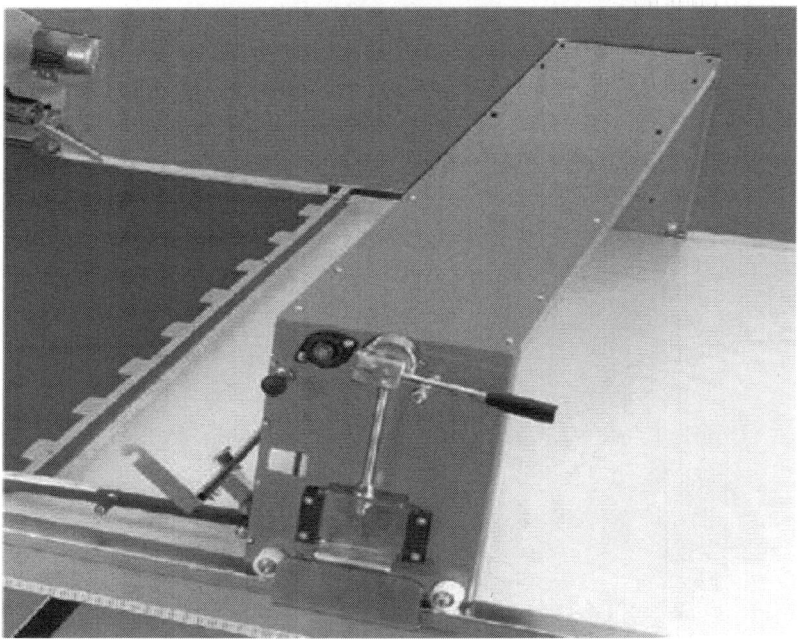

Figure 1.14   End catcher set up at the end of the spread.

The main objective of the end catcher is to hold the fabric spread at the edge by aligning the edge. This will be used specifically in the case of light weight and slippery synthetic material cutting.

- The specific end treatment equipment needed depends on whether the spreading mode is face-to-face or face-one-way. A face-to-face spread utilizes an end catcher and folding blade that work together. These are mechanical parts mounted at opposite ends of the marker to catch and hold the fabric as the blade shapes and creases the fold. An overfeed device may be built into the spreading unit, which automatically feeds extra material when a fold is to be made. End treatments have a major impact on fabric waste. There must be enough fabric at the end of a lay to retain it in place, but any fabric beyond the end of the marker is wasted.

- For F/O/W spreads, a knife box is needed along with an end catcher. A knife box contains a cutting unit (usually a small rotary knife) that operates in a track and cuts across the fabric width when engaged. With face-one-way spreads, each ply must be cut from the roll at the end of the marker. The catcher simply holds the fabric end in place for cutting. As multiple plies are spread, the fold blade and/or knife box must be elevated to the height of the top ply in order to fold or cut the fabric.

6. **Spreading table**

The spreading machine table is another important factor which has the direct influence on the quality of the spread.

Based on the type of fabric, woven, knitted, synthetic fibre material, natural fibre and fabric characteristics different types of tables were preferred for spreading. They are shown in Figure 1.15.

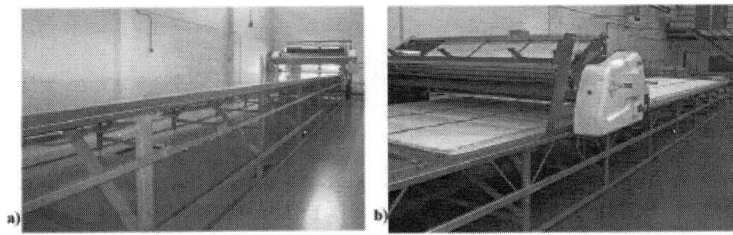

Figure 1.15   (a) Plain spreading table and (b) table with air flotation.

- Smooth surface table.
- Table with air flotation: Air is forced out under the lay permitting it to float on a cushion of air much like a hovercraft. This facilitates moving either a block or the entire spread down the table, when used in conjunction with automatic cutting system.
- The vacuum table: It is a revolutionary technology used in conjunction with servo cutting machines. Replacing the use of cloth weights to keep the fabric in place during cutting, clear Mylar plastic is spread over the entire lay after the marker is in place. Through small holes in the cutting table surface, air is sucked out of the lay. This compresses the lay and stabilizes it.

7. Control panel

An interactive control panel is a graphical device which is used to feed the requirements of cutting room to the machine. It is used to set up parameters and to programme the spreading process as given in Figure 1.16. The main parameters are:

- The lay length,
- The number of plies,
- Start point of the spread,
- Spreading mode,
- Fabric tension,
- Spreading speed,
- Dead head speed,
- Fabric cutting speed,
- Frequency of knife sharpening and
- Fabric cutting length.

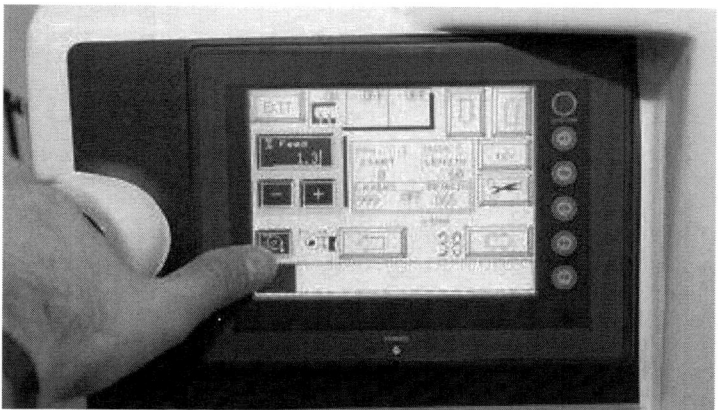

Figure 1.16   Control pannel for automatic spreader machine.

8. Safety system

- The spreading machine is equipped with safety switches like emergency stop buttons (several buttons based on the table length) in the working area.

- The operator platform and spreader head contains an emergency stop button and other sides of the table also.

- The machine is also equipped with sensors like, if any obstacles in the spreading head travel path, it automatically turns of the machine.

- The machines will be turned off if the machine is kept idle for certain predefined time without fabric feed.

Special features of advanced spreading machine

- Cradle with fully automated lifting for ergonomic loading process.

- Programming possibilities for many spreading steps.

- Photocell electric for the examination of the edge control even for difficult materials.

- Dynamic speed control.

- Latest machines can handle heavy weight fabric rolls up to 300 kg.

- Improperly rolled with different tensions can be also handled.

- Automatic adjusting of the tension feeding optional.

- Programmable lay length and end allowance settings.

- Production control software to control the metres, time and percentages, to be able to improve the productivity.

- Based on the type of material the spreading tension can be adjusted from zero as required.

1.3 Cutting

To fabric cut out pattern pieces of garment components as per exact dimension of the patterns from a fabric lay is called fabric cutting. It is totally different from general cutting in which exact dimension is not taken into account. The term fabric cutting is only applicable for garments manufacturing technology.

1.3.1 Requirements for fabric cutting

The following points must be fulfilled in fabric cutting:

1. **Precision of fabric cutting:** Fabric cutting should be done accurately as per exact dimension of the pattern pieces in the marker. Accurate cutting depends on methods of cutting and marker planning. If manual cutting method is used, then cutting accuracy depends on sharpness of knife, skill of operator, and attentiveness of operator. Computer controlled cutting and die cutting have their self cutting accuracy.

2. **Consistent cutting:** Whatever be the cutting method used for fabric lay cutting, it should be ensured that the shape of the cut components from top to bottom lay are of exact size and shape, otherwise the garments produced will be defective.

3. **Infused edge:** During fabric cutting, the friction between the fabric and the blade produces temperature in the blade; the temperature may be up to 300°C. If the fabric contains synthetic fibres, e.g. nylon, polyester, acrylic or their blends, then fused edge may result in the fabric. As because most of those fibres melt at around 250°C. Therefore, sticking of cut edge of fabric will increase the fabric wastage. Moreover, the fused edge after cooling will form hard bid, which will be a problem of irritation during use of garments. To avoid the problem of fused edge formation, the following steps may be taken:

(i)   Reduce the height of the lay,

(ii)   Reduce the cutting speed,

(iii)   Use anti-fusion paper in the lay at regular interval,

(iv)   Lubricate the knife during cutting.

4. **Supporting of the lay:** Surface of the cutting table depends on methods of fabric cutting. The table surface should be capable to support the lay as well as to ensure that all the plies are cut at a time during fabric cutting.

1.4   Methods of fabric cutting

Figure 1.17 explains the different methods of cutting machine used in the apparel industry in details.

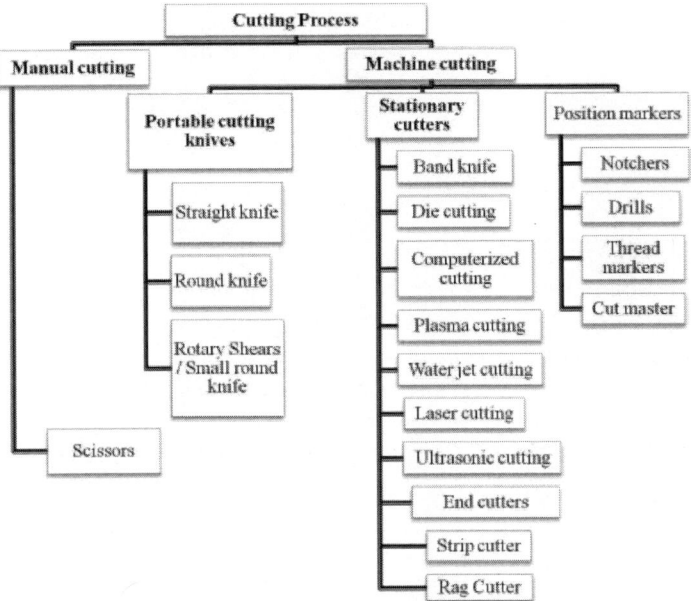

Figure 1.17   Classification of cutting machine.

1.4.1 Hand cutting/manual cutting

Hand shears are used when cutting samples and limited quantities of garments. The cutter must control the shears keeping the cut edge layer adds to the difficulty of accurate cutting, the patterns are often traced in tailors chalk on the top layer of fabric. Hand shears are limited to the cutters physical strength, but usually no more than two layers of fabric due to the loss of accuracy as the shears lift the fabric off the cutting table. This method is slow and unproductive. The process is shown in Figure 1.18.

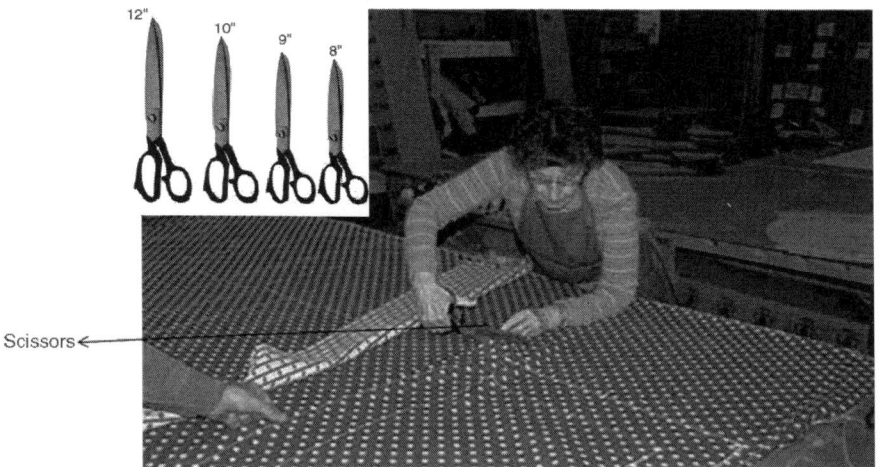

Figure 1.18   Manual cutting process and scissors.

1.4.2 Machine cutting

1.4.2.1 Straight knife cutting machine

The straight knife cutting machine consists of a base plate, an upright stand to hold the vertical blade, motor, a handle for moving assembly, a sharpening device and a handle to transfer the whole assembly from one place to another.

Two kinds of power are required to operate a straight knife.

•   Motor power drives the reciprocating blade.

•   Operator power drives the knife through the lay.

The most important consideration in selecting a straight knife is the power required from the operator to move the knife is the power required for the operator to move the knife through the lay.

The motor power needed is determined by:

- Height of the lay.

- The construction of the fabric.

- The curvature of the line being cut.

- The stroke of the blade.

The greater the power of the motor the heavier will be the machine. The taller the stand, the thicker will be its cross-section and the greater its width, adding resistance to the forward movement on a curve. The greater the blade movement the faster the blade cuts the fabric and the more rapidly and easily the operator can push the machine.

Operator effort is affected by the

- Weight of the motor,

- The shape of the stand,

- Handle height and stroke,

- Sharpness of the blade,

- Effect of the base plate roller,

- Table surface,

- Based on maintenance.

The elements of a straight knife consist of:

(a) A base plate – usually in rollers for easy movement.

(b) An electric motor.

(c) Handle – for the cutter to direct the blade.

(d) Knife (reciprocating motion).

(e) Knife guard.

(f) Grinding wheel – used to sharp the knife during cutting.

(g) Stand.

(h) Roller wheel – to move the machine over cutting table easily.

The detailed parts of the straight knife cutting machine are shown in Figure 1.19 and the functions of the parts are explained below.

Blade

- Normally the available blade heights vary from 10 cm to 33 cm.

- The available strokes vary from 2.5 cm to 4.5 cm. The greater the blade movement the faster the blade cuts the fabric and more easily the operator can move the machine.

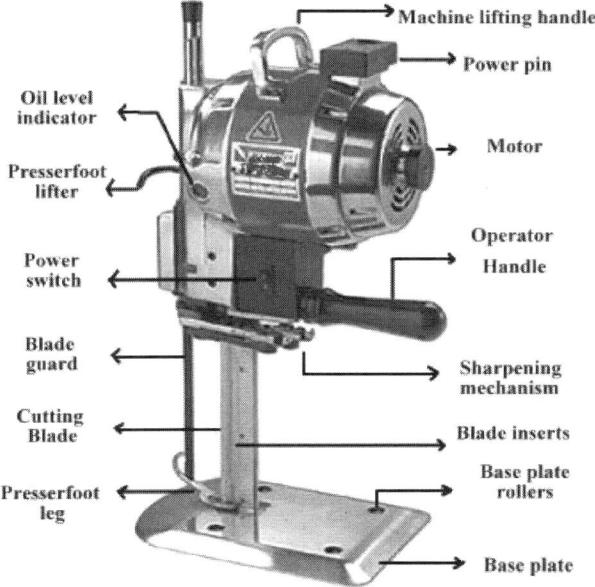

Figure 1.19   Different parts of straight knife cutting machine.

- The normal blade has a straight edge that varies from coarse to fine depending upon the type of fabric being cut. Wavy edged knifes are used to reduce the heat generation and hence can be used for cutting synthetic materials without fusing difficulties. The speed of the blades can also be adjusted by having variable speed mechanism

Classification of blades

The cutting blades for straight knife cutting machine comes in three grades and also in several shapes as shown in Figure 1.20 for the cutting of unusual or difficult material. The grades are:

1. High speed steel – Most popular blade as it wears well, retaining its cutting edge for a long time.

2. Carbon alloy steel – A quality blade less durable than high speed steel but less costly.

3. Special alloy – It retains cutting edge extremely well but is very expensive. Used only based on the requirements. For cutting materials like fibre glass and canvas.

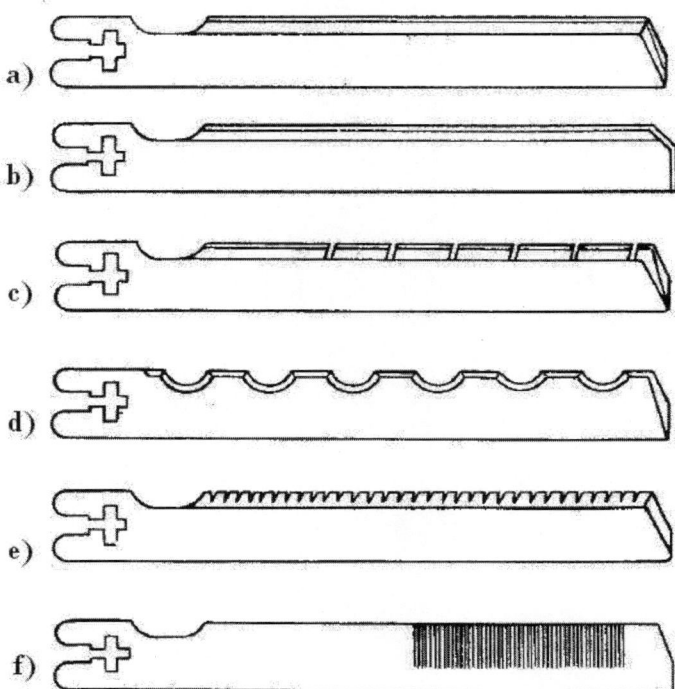

Figure 1.20   Different types of straight knife cutting machine blades.

The various shapes of the blades and their uses are:

(a) Regular blade – recommended for general purpose cutting. Used for cutting softer goods: cotton, wool and knit fabrics

(b) Long blade – 1/8 in. longer with different shaped bottom corner than regular blades. Generally used in the case of loosely woven material or very hard materials like terry cloth, quilting, denims, etc.

(c) Slotted blade – For synthetic leather, supported fabrics, rubber backed fabrics and certain types of plastics.

(d) Wave blade – popularly used for plastics, taffeta and buckram. Designed for materials that will fuse: nylon and other synthetic goods.

(e) Saw blade – used for rubberized fabric, canvas and crinoline.

(f) Serrated blade – for fabrics with designs adhering to surface.

Sharpener

It is small band with grinder/sharpening material on it and its main function is to sharpen the blade. It is necessary to stop the work process during sharpening. The frequency of sharpening depends on the material being cut and on the knife blade.

Motor

- Motor rpm is 3000–4000 and its main function is to move the blade up and down and also slightly in front direction to create a stroke on fabric.
- Knife cuts the fabric very fast due to high speed of motor. That increases the risk of fabric damage.

Fabric presser foot

- It is a device in the machine which holds the layers of the fabric tight so that it will easy for cutter to cut the fabric.
- A presser foot is used to ensure compression of the spread directly in front of the knife and to decrease displacement of fabric plies during cutting.
- It also prevents any loosening of the plies during the return movement of the knife.
- The height of the presser foot is changed according to the height of the spread.

Base plate

- A base plate with the lowest possible profile is needed to ensure the stability of the machine. This facilitates its manoeuver ability and minimizes the risk of deforming the material plies during the work process.
- Independently moving rollers are fixed under the base plate to ensure easy movement of the machine. Wheels are under the base plate to move the machine smoothly.
- Machine weight is around 12–15 kg.

Moving handle

- This handle is to hold the machine as well to move in predefined direction.
- The straight knife is a common means of cutting lays in conventional cutting rooms because it is versatile, portable, cheaper than a band knife and easy to maintain. Even if a band knife is used for main cutting operation, a straight knife will be used to separate the lay into sections for easier handling.

Working mechanism

- The straight knife cutting machine is generally used in the small scale industries. Where the whole spread is cut by the straight knife machine. In the case of large scale industries, the straight knife is used to block the spread before moving to the band knife cutting machine.

- The machine operated by manual power. Here, the machine is pushed or pulled in to the fabric spread by the operator as shown in Figure 1.21.

- Hence the quality of the cutting totally depends upon the operator skill. By pressing the power button operator can operate the machine. After a certain time, based on the type of the cutting material the operator will sharpen the knife frequently.

- The rotary motion of the motor in the cutting machine was converted into the reciprocating motion of the knife by a cam.

Figure 1.21   Straight knife cutting process.

Vertical cutting velocity

The vertical cutting velocity of the straight knife cutting machine is calculated by multiplying the cutting machine motor rpm and cutting knife stroke length.

It can be represented as follows:

$$\text{Vertical cutting velocity (m/s)} \quad \frac{\text{Motor rpm} \times \text{Knife stroke length in cm}}{100 \times 60}$$

If the cutting machine motor rpm is 2850 and the knife stroke length is 7.6 cm

$$\text{Vertical cutting velocity} = \frac{2850 \times 7.6}{100 \times 60}$$

Vertical cutting velocity = 3.61 m/s

The common application areas are:

1. Suitable for mass trimming cotton, woolen, linen leather and chemical fibre goods, etc.

2. Neat cutting, small curvature radius curvilinear cutting.

3. Low noise, stable running, easy to operate and high efficiency.

4. Incorporated with an auto knife grinding device, easier to operate.

Advantages

- Comparatively cheap.

- Can be easily transferred from one place to another place (portable and versatile).

- Round corners can be cut more precisely than even round knife.

- Production speed is very good as up to 10 heights can be cut at a time.

- Garment components can directly be separated from fabric lays.

- Fabric can be cut from any angle.

- Easy maintenance.

- Mainly used for separating the lay into sections easily.

Disadvantages

- Knife deflection is high in risk, when any height is too high.

- Deep curves cannot be cut with the straight knife cutting machine.

- The space between the patterns must be kept relatively high during the marker making and planning for the easy movement of knife width in cure areas.

- Make only lateral cuts into a spread.

- Cannot be used to cut out areas from the centre of garment parts.

- To remove a section of fabric without cutting into the areas requires a slasher.

- Sometimes deflecting may occur due to the weight of the motor.

Precaution to avoid blade deflection:

- Reducing lay height.

- The weight of the motor should be light.

- The operator should be skilled and conscious.

Common settings for straight knife cutting machine

1.   To sharpen the blade

The machine should be clear from lay and the presser foot should be down on base plate. With motor running, move the tripper handle to right (facing sharpener) until it latches and release it. The sharpener will sharpen the entire blade once.

2.   To change the flexibands

After disconnecting the power, push the block 'A' to release right upper flexiband B. Remove the flexiband from there and also from drive C in Figure 1.22. Fix new flexiband over pulley C and between band plate D and blade E then push block A.

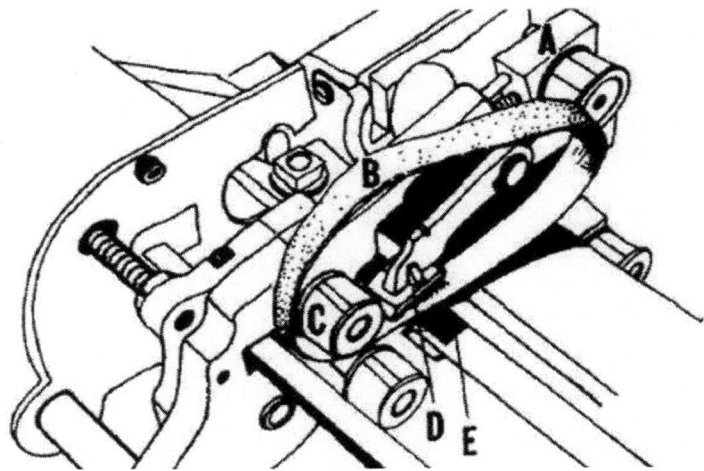

Figure 1.22   Flexi band sharpening set up in straight knife machine.

3.   To change the blade

- After disconnecting the power, lower the presser foot leg to the base plate and lay the machine on left side.

- By pressing the thumb wheel A, move blade to its lowest position. With the knife key B, turn the nut C counter clock wise to unlock the blade. After this raise the presser foot leg and slide the blade D downwards and out (Figure 1.23).

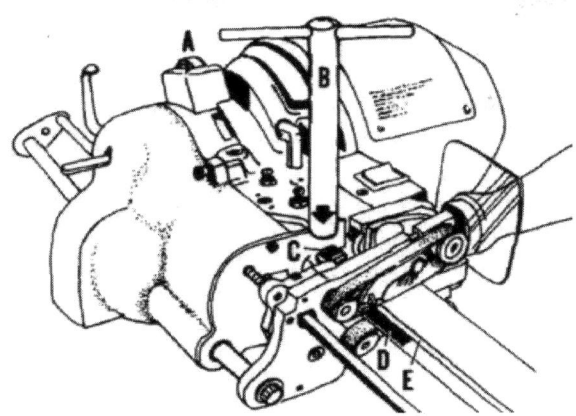

Figure 1.23 Knife set up assembly in straight knife cutting machine.

- The new blade can be fixed up against the locking bolt.

Maintenance for Straight knife cutting machine

Oiling

- Oil the machine daily. Before starting the machine check it and make sure that the oil cup is half filled all time.
- Use good grade oil and do not use sewing machine oil or any other light oil as it is not designed to give the lubrication needed for the cutting machine.

Cleaning of inserts

- The inserts which guide the blade in the stand must be cleaned occasionally and whenever the blade is changed.
- This will avoid cocking of blade because of dirt accumulation. This ensures that the blade being sharpened as planned every time.
- Clean the inserts by sliding the saw tooth slot cleaner up and down the back of the inserts.

To clean sharpeners

- Run the sharpener to its lowest position and turn off the motor. Blow out the lints and dust with compressed air.
- The gears must be checked for dirt or bits of cloth under the gears carrier block.

Motor cleaning

- Over a period of time the dust and lint will build up inside the motor and prevent proper cooling.

- To clear this with the motor in running condition point a stream of compressed air into the back housing and then into the front housing to eject dusts and lints.

Base plate roller cleaning

- The rollers in the base plate do not roll freely, since the dusts are accumulated. To blow out the dust or dirt in the rollers, air stream is used.
- Oiling the roller will collect dust and dirt and cause the rollers to bind.
- Use a powdered graphite for lubrication if necessary.

Trouble shooting

1. Blade edge not sharp
   - Check for worn-out blade
   - Check for dirt and worn out flexi-bands
   - Check for misalignment of standard and presser foot leg

2. One side of the blade not sharpening
   - Check for a weak or broken guide spring
   - Check for worn inserts
   - Check the positions of blade safety arm

3. Blade sharpening at an angle
   - Check the inserts for dirt
   - Make certain that the blade is tight against the back of the inserts
   - Check the misalignment of presser foot leg and standard
   - Check for worn presser foot leg guide or weak presser foot leg guide spring

4. Bottom of the blade chewed out
   - Check the position of the blade safety arm
   - Check the vertical movements of the blade carrier

5. Sharpener not running smoothly
   - Check the molded pulley for excessive wear
   - Oil gears under carrier block and on upper gear block

6. Sharpener fails to operate
   - Make certain that the presser foot leg is down completely

- Check for cloth jammed in the gears
- Check for broken teeth of gears

7. Presser foot leg slipping/ fails to operate

- Weak or broken presser foot trigger spring
- Check for worn V block in front plate
- The presser foot leg fails to operate if the sharpener is not all the way up
- Or even after sharpener is all the way up, the safety lock screw interferes with the movement of presser foot leg

8. Motor slow in reaching full speed

- Check the proper voltage in being delivered in the line
- Fuse probably out on one line
- Ground wire incorrectly connected to the machine

9. Motor rotates in wrong way/or not starts

- In single phase – the back house incorrectly aligned with the stator
- In three phase – one electrical line may not work
- Check the connector is firmly attached to the pins
- Defective switch or shunt wire

10. Machine does not roll freely on table

- Clean the rollers in carrier
- Condition of the surface of the cutting table
- Cushion in roller carrier worn causing base plate to drag in table

1.4.2.2 Round knife cutting machine

This machine is called round knife cutting machine because its cutter is round but slightly octagonal in shape. This machine is small in size, flexible and used for small production. It is also a popular cutting machine, shown in Figure 1.24.

Working of round knife

- As mentioned in the straight knife cutting machine, the round knife also operated in the manual power. Hence, the cutting efficiency totally depends upon the operator skill.
- During the cutting process, the operator operates the machine by holding the table of the machine either by pushing or by pulling it through the fabric spread.

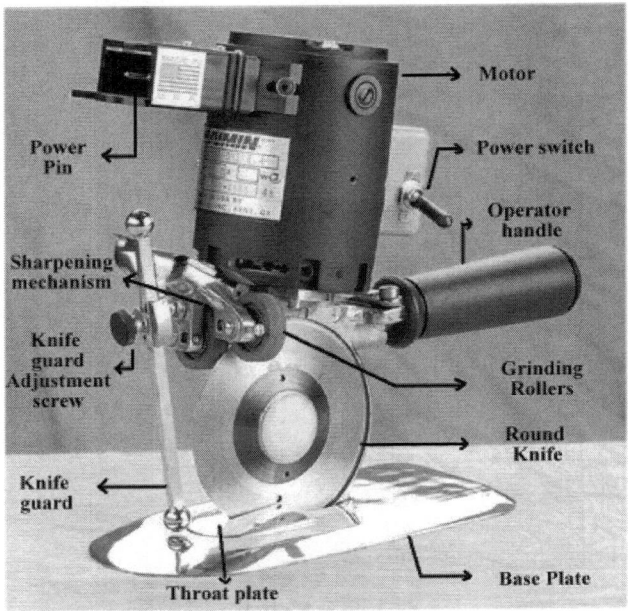

Figure 1.24   Round Knife Cutting Machine.

- During this process, the thin base plate goes underneath the cloth for easy and precise cutting.

- The rotary movement of the motor is directly transferred to the cutting knife. Instead of reciprocating motion the knife rotates continuously in the round knife.

- A sharpening mechanism is fixed on both the side of the knife to sharpen the knife while required.

- The rotating circular knife cuts the fabric according to the pattern.

- The cutting depth of the circular knife is limited and hence, the machine is used in the case of low number of plied spreads and also for lesser order quantities.

Base plate

- Supporting foundation to balance the cutting.

- Vary in shape, sizes depending upon the size and weight of the knife.

- Guides the knife in relation with the cutting table.

- They are supported by bearing rollers.

- Edge of the plate is sloped.

Power system

- Controls the motor and potential cutting speed.
- The amount of power needed to cut a spread depends on the height of the spread and the density of the fabric to be cut.
- Horsepower determines the amount of thrust.
- Higher speed means knife moving faster.
- Greater horsepower increases machine power and weight also.

Sharpening device

- Appropriate for the specific blade types.
- Blades dull quickly when cutting a deep spread or dense fabric.
- Prevents rough, frayed or fused edges.
- May be stone or emery wheels or abrasive sharpeners.

There are two types of sharpening devices available for the round knife machine as in Figure 1.25.

Figure 1.25 Grinding rollers for round knife cutting machine.

1. Gliding grinder – most commonly used method, where the grinding stones are set perpendicular to the knife. It sharpens the cutting edge faster.

2. Normal grinder – uses slightly large stones which are almost parallel to the knife. This gives smoother cutting edges for cutting sheer fabrics, synthetics and plastics.

Cutting knives

Cutting knives come into two different grades for round knife cutting machine.

1. Carbon steel – commonly used for the cutting of all materials.

2.   High speed steel – used for the heavy materials like canvas, which will quickly dull the carbon steel kind of knife.

The round knife also available in different shapes other than the round shape as shown in Figure 1.26.

Figure 1.26   Different knife shapes for round knife cutting machine.

Handle

To grip, guide and propel the knife through the spread.

Blade guard

•   Rests on the top ply to help stabilize the spread and to protect the operator's hand.

•   Metal mesh gloves are also available as a safety device.

Special features

•   It contains a round but slightly octagonal type knife with sharp edge.

•   Knife diameter varies from 6 to 20 cm.

•   Motor rpm is 800–1600. It depends on machine.

•   A handle for the cutter to direct the knife.

•   Easy to handle and movement due to low weight.

•   Knife is lubricating manually.

•   Three types of knife edge can be used for cutting different objects. Such as waved edge, toothed edge and circular edge.

•   The most widely used machines are those with 100 mm and 110 mm knives. Blades of different shapes are used for cutting different materials.

•   Round blades are used to cut light fabrics but polygonal blades with 4, 6, 8 and 10 sides are used to cut thicker and harder materials.

•   A round knife rotating so that the leading edge cuts downwards into the fabric.

- Flexible movement helps to cut non-linear shape.
- Base plate gives support for fabric.
- Maximum 40% of the knife diameter can be used for fabric lay.

Advantage of round knife cutting machine

1. Suitable for cutting single ply as well as multilayer (say 20–30 layers).
2. Easy to handle and operate.
3. Suitable for small scale cutting.
4. Round knife cutters are lighter (between 3 and 11 kg) than straight knife machines and are therefore easier to move.
5. Suitable for gentle curve line cutting. To cut the larger part of the garments.
6. With the same speed, its efficiency is 10 times greater than the straight knife.
7. Round knife machines are the most effective for cutting slippery materials as the rotary movement of the knife ensures the continuous compression of the fabric plies.

Disadvantage of round knife cutting machine

1. Very low speed and knife height.
2. Manual grinder is used.
3. Low productivity since few number of lay can be cut.
4. Difficult to cut small components and high curve line. Not suitable for cutting curved lines in high lays – the blade does not strike all the plies at the same point. Therefore, round knife only used for lower lays of relatively few plies.
5. Not suitable for large production.
6. Lubrication is manually done.

Maintenance

Cleaning of sharpening mechanism

- After repeated uses, the sharpening stones become coated with grease and dirt and do not sharpen the knife effectively.
- To remove this coating, stone cleaners can be used on the stones or else, if the sharpening device works based on sand paper belt, it can be replaced frequently.

Knife maintenance

- The blade should be replaced once it wears out around 5 cm from its diameter.

- Unscrew the knife lock, which is located in the knife centre, by rotating counterclockwise with the knife key and remove the blade.

- Lift the old blade from gear and install new one, by facing the printed letters or logo facing outside, rotate the gear manually and make sure the blade is fixed perfect. Secure the lock.

Gear lubrication

- The gears and warm ring, which is located behind the knife blade, need to be lubricated by grease after every week.

Throat plate adjustment

- As the machine works continuously, the knife wears. This increases the space between the base plate and knife. This should be adjusted every time to improve the cutting accuracy and quality.

- The space between the edge of the knife and the front of the slot of the throat plate is 1.6 mm.

- The knife should be in the middle of the throat plate slot.

- The throat plate can be moved front and back by loosening the screws at the bottom of the base plate.

Motor cleaning

- The dust and lint accumulation in the motor should be cleaned after continuous working.

- This will reduce the unwanted heating up of motor during the working. Use a point stream of compressed air into the top of the motor and then in the side and edge motor housing.

Base plate roller cleaning

- If the rollers in the base plate do not roll freely, blow out the dust and lint or dirt in the roller by using a stream of compressed air.

- Do not oil the roller; it will cause the dirt or lint accumulation to bind the rollers from rolling.

Over all machine maintenance

- The motor should be blown out with compressed air occasionally to remove the dust and lint for more efficient and cooler operation.

- If the machine stops or slows down from the fragment of material collecting in the gear, remove the blade and then the gear, clean the cloth fragments from the gear and standard and replace the blade.

Trouble shooting in round knife machine

If the machine does not operate

- Check that the current connector is firmly attached to the terminal pins and the machine is connected to the terminal outlets.

- Check for worn carbon brushes (fuse).

- Check for the dirt or cloth jammed behind the blade or in the gear.

1.4.2.3 *Rotary shears/small round knife cutting machine*

Special small sized round knife cutting machines are available for cutting single or multiple plies of material (for spreads up to 10 mm), shown in Figure 1.27.

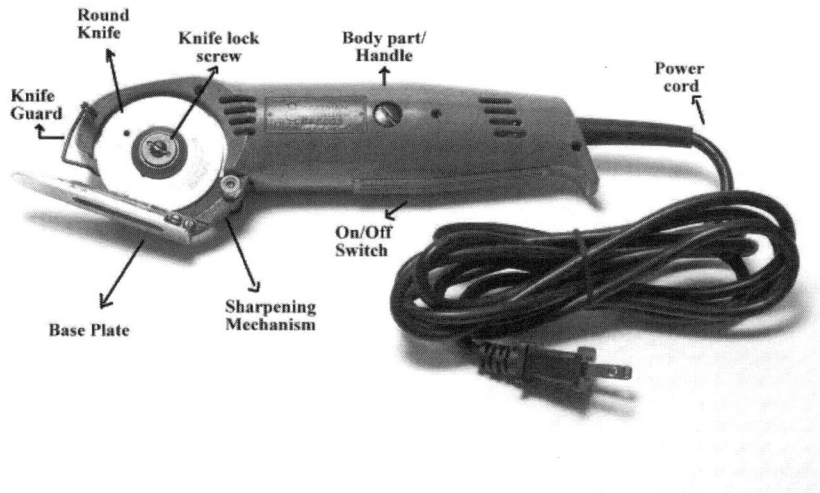

Figure 1.27   Parts of small round knife cutting machine.

- These differ in shape, have a small diameter knife (typically 50 mm) and are light in weight (0.5–1.5 kg). The machines may be mains or battery powered.

- Smallest and most widely used rotary shears are perfect upgrade to or replacement for, manual shears. Streamlined, lightweight construction helps eliminate operator fatigue.

- Built-in sharpener mechanism for ease of use.

- Equipped with round blade for general use, or optionally with a hexagon or four-sided blade for difficult-to-cut materials.

Parts and Function

Knife – Generally round knife used to cut the fabric. Based on the type of application variations in the knife like hexagonal can be preferred.

Knife guard – A small metal rod in front of the knife to avoid direct contact of hands with the knife during the cutting process

Lock screw – Knife lock screw is the essential part, which holds the knife in position. The screw needs to be removed while changing the knife every time.

Carbon bushes – Carbon bushes are inside the body and motor cases. The bushes need to be replaced after every 200–500 h of used based on the manufacturers.

Sharpening mechanism – Knife sharpening mechanism is on the both sides of the blade. The mechanism need to be cleaned after every usage. Dust accumulation in the device will reduce the sharpening efficiency and hence the cutter efficiency.

Base plate – The base plate is the plate which moves under the fabric plies during the cutting process. The base plates can be removed and replaced with cutting guides sometime.

Body/handle – The machine body is a plastic outer cover of electrical motor assembly which is also used as a handle to operate the machine. But machines are available with long handles too. The working of small round knife cutting machine is shown in Figure 1.28.

Figure 1.28   Fabric cutting using small round knife cutting machine.

1.4.3 Stationary cutting machines

1.4.3.1 Band knifes

Use of band knives

- Band knives are used where higher standard of accuracy is required.

- As a first step, by using straight knife the lay will be sectioned that by using band knife, the accurate cutting will be done.

- When smart parts like collars, pockets are cut, a template of metals is used in the shape of pattern to get correct shape.

- Mostly used in men's wear. Generally used to cut large parts of garments such as jackets and overcoats (Figure 1.29).

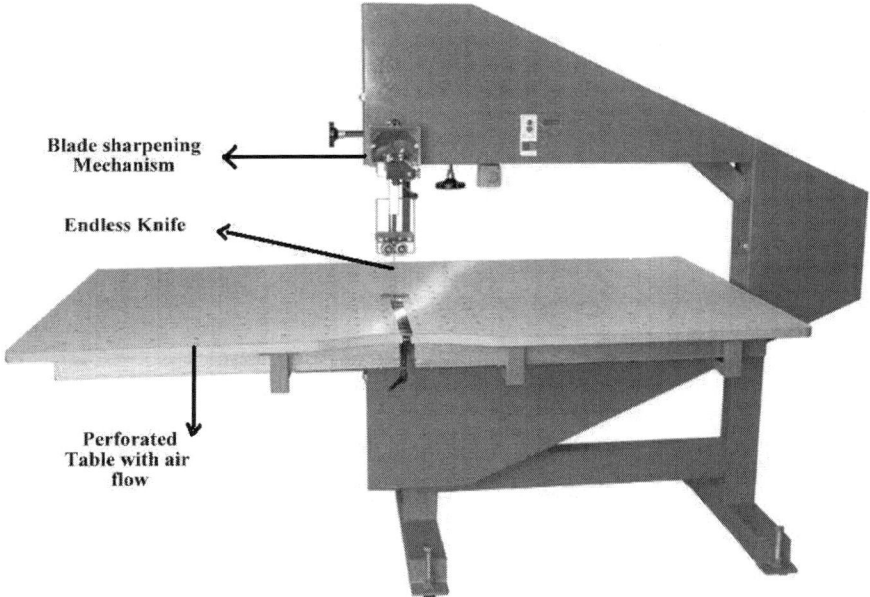

Figure 1.29   Band knife cutting machine

Parts of machine

Knife

- The length of the band knife may differ based on the application requirement from 2825 mm to 4920 mm.

- The longer it is, the larger and more powerful the cutting machine will be.

- The width of the band knife may be 10, 12, 15, 20 or 35 mm. Knives of 10, 15 and 20 mm width are the most widely used. A narrower knife has a smaller cutting surface; it has a greater degree of precision, although the steel band will become blunt more quickly.

- The thickness of the knife does not differ as widely (0.45 mm, 0.50 mm). The standard thickness is 0.45 mm. The path way of the knife in cutting machine is shown in Figure 1.30.

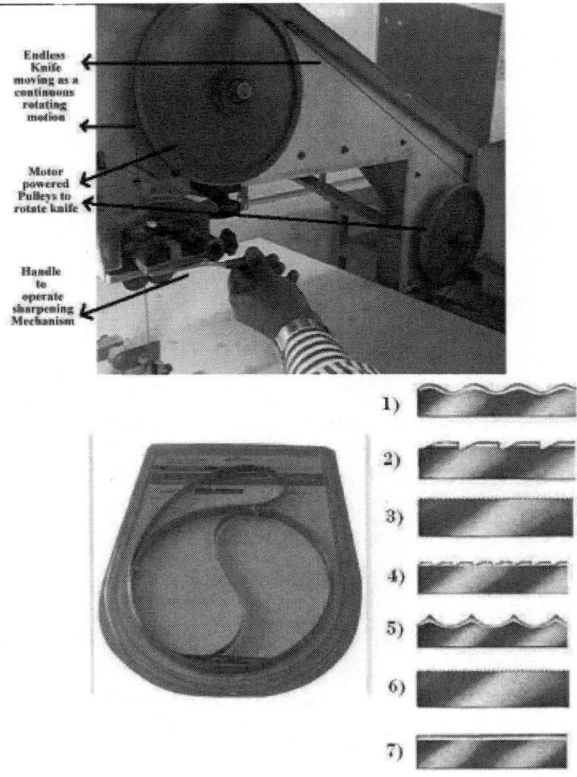

Figure 1.30   (a) Knife path way of band knife cutting machine and
(b) Types of Band knifes.

Types of knives

1. Wavy edge – Wavy edge band knife for cutting materials such as tissue paper and textiles, the wavy edge gives a clean, smooth finish. Available in widths from 13 mm to 25 mm and in a short wave pitch or long wave pitch.

2. For cutting soft plastics.

3. For cutting the most wide-ranging types of rubber.

4. For cutting hard plastics.

5. Scallop edge band knife for cutting harder synthetics-based textiles. Also used for cutting rubber seals and gaskets, packaging material, and other applications where an initial penetration is needed. Available in widths from 10 mm to 25 mm with various pitches and gullet depths.

6. Diamond toothed edges for cutting the most wide-ranging types of rubber.

7. Straight edge – Straight edge band knife is available with a single or double edge, for cutting flexible foam or other soft and fibrous materials. Available in widths from 3 mm to 85 mm for vertical and horizontal cutting and splitting, with a variety of bevelling options.

Working mechanism

*   Three or more pulleys, powered by electric motor, with continuous rotating blade mounted on them. In band knife the knife is in a fixed position. The fabric will be moved. The principle is opposite to straight and round knife. Figures 1.31 and 1.32 show the sharpeneing rollers and cutting head of band knife cutting machine.

Figure 1.31   Sharpeneing rollers for band knife cutting machine.

*   The operator either pulls or pushes the section of lays. In this method machine is stationary but fabric is movable. Band knife cutting machine is look like a wood cutter machine. Band knife is used for precision cutting small parts of garment. This band knife cutting machine has

lower noise, less temperature up, high power, less vibration and auto sharpening device.

- This machine with special blower decreases resistance between fabrics and table, which enables the fabrics be moved easily and be cut precisely. Air blower blows the air to minimize the weight of fabric. Balls in air blower help to move the fabric in different direction.

- And it could adjust the speed to fit different fabrics. High speed motor is used.

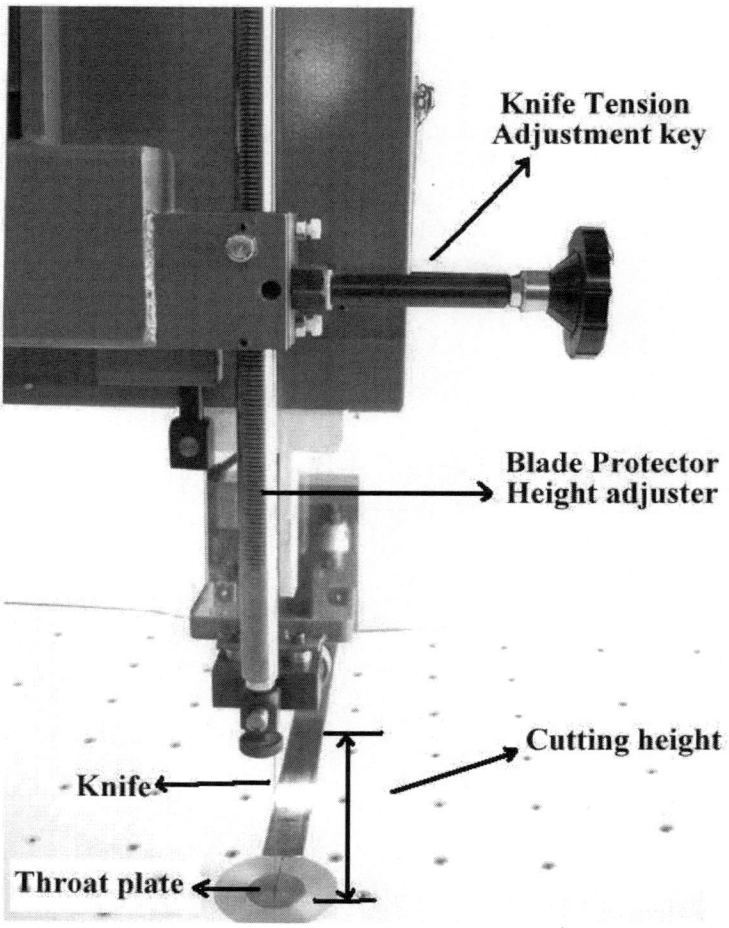

Figure 1.32   Cutting head of band knife cutting machine.

- Requires a large space for it. Blade moves vertically through a flat working table.

- It comprises a series of three or more pulleys, which provide the continuous rotating motion of the knife. An endless knife is used here. Knife is usually narrower than on a straight knife.

  - Automatic grinder is used to sharpen the knife. A large size of table is used to support the fabric and for cutting.

  - The cutting machines operate with either a fixed knife speed (e.g. 14 m/s, 16 m/s) or an adjustable speed which accommodates a wider range of fabrics (8–18 m/s, up to 30 m/s).

Band knife cutting machines have been recognized around the world for many years for their high quality even when faced with difficult customer-related conditions and in applications with demanding materials. The working of band knife cutting machine is shown in Figure 1.33.

Figure 1.33 Cutting process in band knife cutting machine.

Advantages of band knife cutting machine

- Suitable for any types of line.

- Very large productivity for limited products, such as collars, cuff, placket, etc.

- Automatic grinder grinds the knife instantly.

- Air blower helps to reduce the fabric weight which increases smooth movement of fabric.

- Possible to cut 90° angle of the lay.

- Intensity of accident is low.

Disadvantages of band knife cutting machine

- Not suitable for large component due to the length of the table.
- Work load is high as machine is stationary and fabric is movable.
- Running cost is higher.
- Required fix space.
- Not possible to cut fabric directly.
- Not possible to cut fabric directly from lay.
- Block pieces of fabric required in bundle form to cut by this machine.

1.4.3.2 Computer controlled cutting knives

Figure 1.34 shows computer controlled cutting machine, for high-volume production, computerized apparel cutting offers a wide range of benefits over manual cutting.

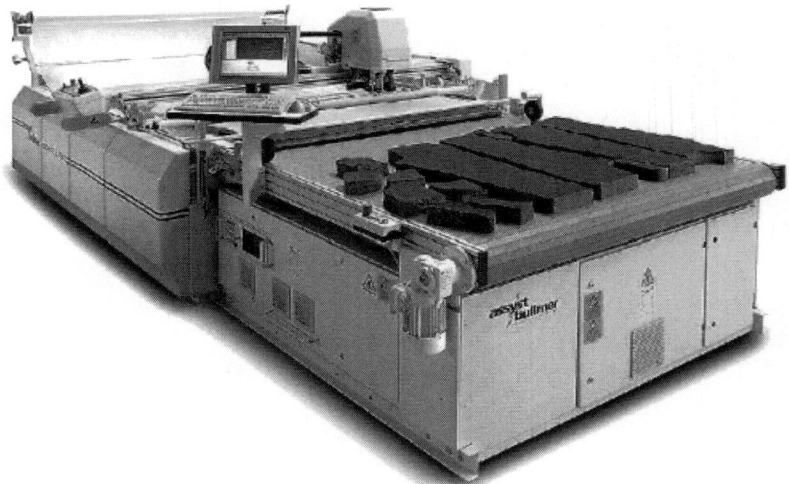

Figure 1.34   Computer controlled cutting machine.

1. Perfect fit every time.
2. Greater yield.
3. More design flexibility.
4. Less cost.

Parts and functions of computer controlled cutting machine

Cutting table

- The cutting surface of the table has nylon bristles, which are flexible. This supports and keeps the lay in place.

- The flexibility of the nylon bristle permits the penetration and movements of the cutting knife as shown in Figure 1.35.

- These bristles allow the passage of air through the table to create vacuum. Thus the created vacuum reduces the height of the lay and hold the fabric lay in place on the table.

Figure 1.35   Nylon bristles in the table surface of computer controlled cutting machine.

Carriage

- This supports the cutting head and has two synchronized motors which drive the machine to and fro.

- The carriage moves on rail tracks laid parallel on the side of the table.

- A third motor positions the cutting head on the table. These two movements are coordinated to give the knife position on any point of the table.

Cutting head

- This consists of knife, automatic sharpener and a motor. The motor rotates and positions the knife to the line of cut on curves and adjusts the pressure on each side of the blade.

- The sharpener sharpens the cutting blade at regular intervals.

- A lift and plunge feature enable the knife to negotiate sharp corners and V-shaped notches.

- A motorized drill behind the cutting head can provide drill holes too when it is required.

Control cabinet

This houses the computer and the electrical components required to drive the cutter, its carriage and vacuum motor on table. The different parts of computer controlled machine are shown in Figure 1.36.

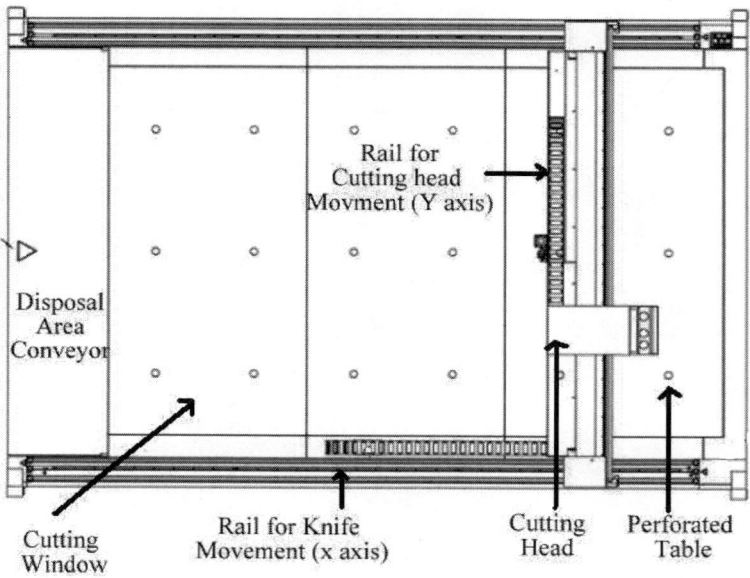

Figure 1.36 Various parts of computer controlled cutting machine.

Spreading and cutting tables

- The cutting process of this computer controlled cutting machine takes very less time than the spreading process and therefore a typical arrangement would consist of four spreading tables supplying to one cutting table.

- Placing the lay on the cutting table, the spreader spreads the lay on a conventional cutting table equipped with air floatation.

- Paper is spread below the bottom ply so that the lay can be moved into the cutting table without distortion.

Polyethylene sheet

After the laying process, a sheet of polyethylene covers the top of the lay which assists the creation of vacuum and allows the significant compression of the lay.

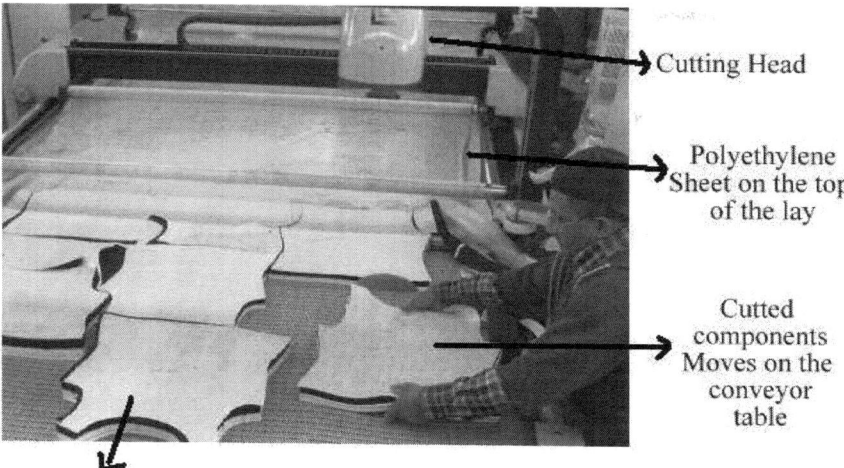

Cutting Head

Polyethylene
Sheet on the top
of the lay

Cutted
components
Moves on the
conveyor
table

Marker Planning sheet on the
Top of the lay

Figure 1.37 Cutting process computer controlled cutting machine.

- This also provides support in bottom plies during the process of cutting and this paper is mostly perforated. This makes this paper to enable the vacuum in compressing the lays without difficulties (Figure 1.37).

Height of the lay

- Different systems are available to cut the different height of the compressed plies.

- The maximum height is around 7.5 cm but however that depends upon the manufacturers.

- The number of plies depends upon the thickness of the individual layer.

Cutting knife

There are different types of knives (Figure 1.38) used in the computer controlled cutting machine. They are:

- *Reciprocating knife* – Oscillating knives are frequently used to cut thick materials, multi-ply spreads or patterns with a complex shape. The oscillating knife is ideal for cutting a wide variety of materials.

- *Drag knife* – The knife blade is angled. The angle of the blade depends upon the properties of the material. During the cutting process, the knife is dragged along the profile of the cut component.

- *Rotary blade knife* – The rotary blade is circular and rolls over the material. It is guided by a motion controller. Rotary blade knives are used to cut plastics.

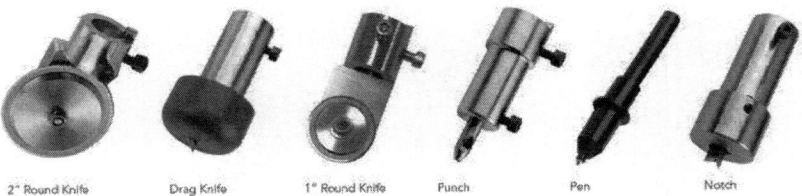

Figure 1.38   Types of cutting knifes in computer controlled cutting machine.

- *Router* – During the cutting process, the cutting tool rotates around its vertical axis and also moves along the profile of the cut component.

Cutting process

- The process of cutting follows the sequential order of the laying the spread on the cutting table and loading the disc in the computer. Most of the computer cutting machine comes with the attachment of automatic spreading machine.

- From the marker making process, the predetermined width and length of the lay were fed to the computer and the spreader spreads the required measurements along with determined number of lays.

- Once the spreading completed, the table conveyor moves the lay to the cutting window. Where, the cutting head receives the instruction from the computer and cuts the lay into pattern pieces.

- After loading the disc into the computer, the operator positions the cutting head into their origin point over the corner of the spread. This provides the computer with a reference point (Figure 1.39).

- The computer controlled knife cuts according to the instruction from the computer rather than by following a pattern line drawn on a marker, it is possible to remove the printed marker plan from the top of the lay.

- However, the marker plan sheet is kept on the top of the lay for the verification of supervisor's reference.

- Care must be taken to load the correct disc because a simple error would lead to a costly wastage.

Special features

- The computer controlled machines feature a highly sophisticated cutting mechanism with any one type of (reciprocating in case of textile materials) knife for precision cutting through fabrics.

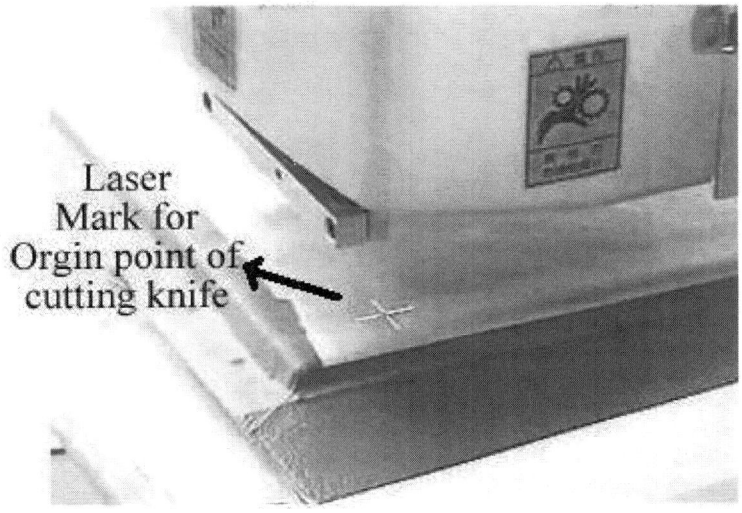

Figure 1.39   Start up marker point in computer controlled cutting machine.

- The latest machinery comes with special function to cool the blade while cutting material that can melt and adhere to the blade in cutting head. An improved knife sharpening system also permits more precise cutting.

- Different manufactures offer different features, however up to a maximum height of 7.5 cm usually when compressed were possible in this cutting machines and hence the number of plies, being dependent on the nature of the fabric.

- The overall speed of cutting garment parts is generally between 5 and 12 m/min and the maximum cutting speed is 30.5 m/min (1200 in./ min.).

- Integrated vacuum system holds materials securely in place for effective and accurate cutting. Integrated "intelligent" zoned vacuum concentrates and holds material at cutting location.

- During the process the knife blade gradually wears from repeated cutting and sharpening, precision cutting becomes increasingly difficult. To alleviate this situation, latest cutting machines from the leading manufacturers measure the knife blade width automatically, thereby maintaining accuracy for such detailed cutting as sharp corners and notches. The knife width auto measurement function also monitors blade wear in real time, notifying the operator when replacement of the blade is necessary.

- The machines also equipped with internal suction chamber which enables cutting waste to be disposed continuously.

- When cutting along a common pattern line that is shared among two patterns that is next to each other, the software designed to cut the line with a single cut. By avoiding repeated cutting, the risk of cutting errors is reduced.

Maintenance and troubleshooting

- Check for water in the pneumatic filter at the rear of the machine. Drain if required. A frequent need to drain this filter is an indication that plant air system is in need of service or repair.

- Check the oil level in the lubricator for the reciprocating knife or pneumatic router. It is located in the middle of the backside of the gantry. Refill as required with the pneumatic oil provided by manufacturer.

- Rotate and/or flip your cutting underlay to equalize its wear.

- Check the air filters in the bottom of the electrical cabinet and clean if required. You must unplug the electrical cord before opening the box.

- Check the vacuum blower for dirt and dust accumulation and clean as required. Be sure to check inside the muffler as well.

- Inspect vacuum holes in the surface and clean any clogged holes. Cover the surface and turn on the vacuum and check for any leaks in the plumbing inside the base. Check the operation of the zone control valves and pneumatic actuators.

- Check and adjust the reciprocating knife/pneumatic router lubricator.

1.4.3.3 Die cutting machines

The die cutting machine works directly opposite to the previously described system wherein other systems, the cutting knife moves fast to cut the fabric. But in this case, the cutting is involved by pressing a rigid blade through the lay of fabric. The die cutting machine for textile cutting is shown in Figure 1.40.

The die (knife) will be in the shape of pattern periphery, including notches. These dies can be of two types:

1. Strip steel type – cannot be re-sharpened. Manufactured by bending a strip of steel to the shape required and welding the joint.

2. Heavier gauge forged dies – can be sharpened but costly.

Figure 1.40   Die cutting machine for textiles.

Die cutting machines are more accurate than any other machine. But because of cost of die, it is used only when required. Die cutting offers much faster cutting than blade cutting machines. More economic for small parts, hence mostly used in cutting of collar, cuff, pockets, bra parts, etc.

*   Die cutting operation involves:

    *   Placement of fabric

    *   Positioning the die on the fabric

    *   Engaging the machine to press the die into the fabric

*   Used mainly for leather, coated and laminated materials.

*   Areas where the same patterns are used over a long period, e.g. collar, pocket flaps.

It is proportionally more economic for small parts, which have a greater periphery in relation to their area than to large parts. The examples of dies were given in Figure 1.41. In addition, the level of accuracy demanded of small parts is not only greater but correspondingly more difficult to achieve with conventional knives.

Die is prepared with metal strip according to the shape of pattern. Die press generally has a cutting die that just penetrates the cutting pad in order that the fibres of the lowest ply are completely separated. For die cutting, the spreader spreads a lay to the required placement of dies. The spread is cut into sections to allow transport to the cutting pad. The gap between two dies 2 and 3 mm is placed when die is taken to cut fabric lay. The process is given in Figure 1.42.

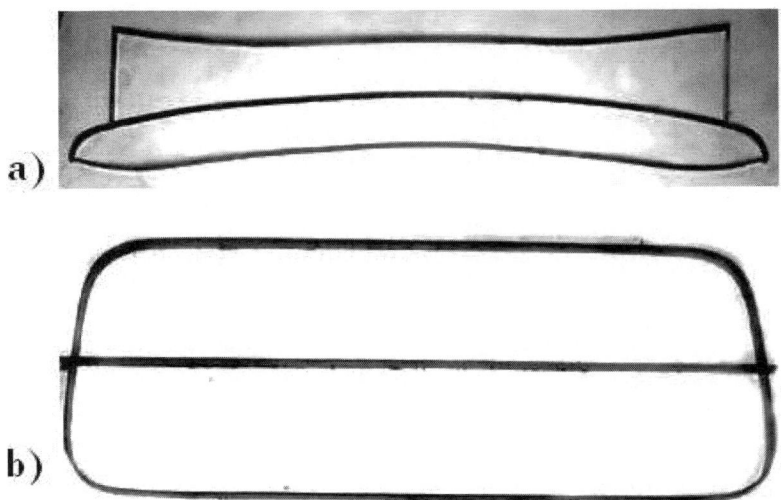

Figure 1.41   Metal Dies in the shape of a) Shirt collar and b) Cuff.

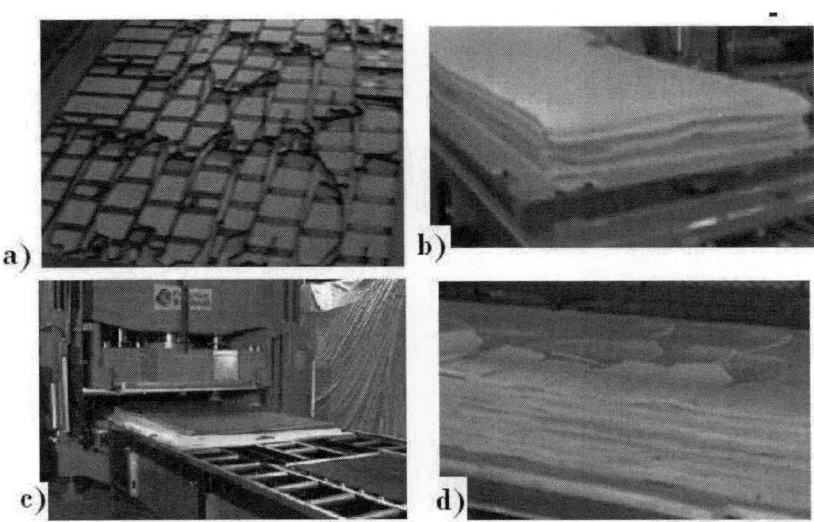

Figure 1.42   Die cutting process)a) Dies are arranged as per marker (b) The
                 fabric is spreaded on the table (c) The table air floatation moves
                 the fabric to the die cutter and (d) Cutted fabrics.

The die press generally has a cutting arm supported by a single pillar at the
back of the machine. It swings to the side to allow the placing of dies on top
of the fabric.

The two types of die press are given below.

1.   Impact – single press on the die.

2.   Hydronic (Hydraulic and Electronic) – which exerts continuous pressure on die till it cuts the fabric.

Working of continuous die cutting machine

The latest advancements in the die cutting machine made the production process as continuous one from the batch process. In this method, the die and press set up unit is stationery and a conveyor belt carries the material to the pressing section. Where, the material is pressed by the die and the cut material is further carried away by the out feed conveyor. Figure 1.43 shows continuous die cutting machine for textile and paper cutting.

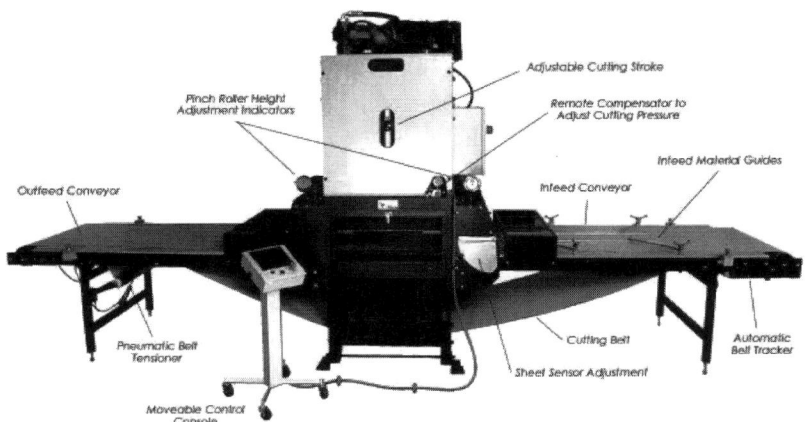

Figure 1.43   Continues die cutting machine for textile and paper cutting.

Parts and functions

Infeed and outfeed conveyor – These belt mechanism works as a material feeder and out feeder.

Cutting stork adjuster – Based on the type of material used, the die press height can be adjusted using this mechanism.

Cutting pressure adjustment indicator – The cutting pressure can be adjusted by using user interface panel. The adjusted pressure will be displayed in the cutting pressure adjustment indicator. The pressure level can be selected based on the type of material used and die selected.

Pinch roller – It helps the fed material to move into the correct height while reaching the pressing section and also hold the textile lay firmly both the sides against the die.

Sensors – Used to detect the material position in the feed conveyor.

Die cutting mainly used for embroidery applique, shoulder pads, cap, shoe and backpack items cutting to get high standard of accuracy but only appropriate to situations where large quantities of the same pattern will be cut. This appliance is very accurate for sharp corners and circular patterns but not suitable for larger parts cutting. But this process may increase production cost, cutting time and higher fabric wastage due to use of block pieces. Operator should be more careful to handle this device because a heavy metal bar creates 20 ton pressure during cutting.

Advantages

- Extensively used to cut sharp corners of small parts of dress accurately.
- Most useful to cut at any shape or any angle.
- Comparatively less time required.
- Best method of cutting knitted fabric.
- Best and cost effective when used in the case of large quantity production.

Disadvantages

- When the die press forces the dies through the fabric it also forces a narrow wedge of fabric between the dies. The narrow wedge exists because the sharpened cutting edge of the die is necessarily of narrower gauge than the top of the die. Thus if dies are butted together, they touch at the top but show a small gap at the level of the cutting edges. The action of the press will compress this narrow wedge of fabric to the point where it will rupture the dies. Hence, it is necessary to leave a significant gap between two dies, say 2–3 mm. Similarly a single die will not cut satisfactorily if placed closer than 3–4 mm to a previously cut edge.
- The capital cost of the large area die press is high.
- There are lots of engineering problem in large lay cutting.
- It is difficult to have dies in the case of frequently changing market.
- Fabric loss is high due to the need of block of fabric lay.
- To change the style quickly is impossible.
- Difficult to cut large components of dress.
- First die set up times.

1.4.3.4 Laser cutting machine

Laser cutting systems combine the heat of the focused beam with assist gas, which is introduced through a nozzle coaxial to the focused beam. The best example of the chemical effect of the assist gas is the use of oxygen for the cutting of steels where performances are increased by the exothermic reaction of combustion of iron in oxygen. Another example is clean cutting stainless steel with high-pressure nitrogen. As the laser beam cuts the stainless steel, the high-pressure nitrogen blows the melted material away.

While carbon dioxide lasers are capable of generating tremendous heat intensity, it is an incorrect assumption that they are capable of vaporizing and cutting all known materials. The question of suitability of using a laser for cutting that material hinges on how well it handles the added energy input. That interaction is dependent upon three key factors of the material.

- Surface condition – how well it initially absorbs the energy.

- Heat flow properties – its coefficients of thermal diffusivity and conductivity.

- Heat phase-change requirements – the amount of excess heat required to induce a change as a function of the materials density, specific heat, and latent heat of vaporization.

Basic principle

The basic mechanism of laser cutting is extremely simple and can be summarized as follows:

A high intensity beam of infrared light is generated by a laser. This beam is focused onto the surface of the workpiece by means of a lens. The focused beam heats the material and establishes a much localized melt (generally smaller than 0.5 mm diameter) throughout the depth of the sheet.

The molten material is ejected from the area by pressurized gas jet acting coaxially with the laser beam as shown in Figure 1.44. With certain materials this gas jet can accelerate the cutting process by doing chemical as well as physical work. For example, carbon or mild steels are generally cut in a jet of pure oxygen. The oxidation process initiated by the laser heating generates its own heat and this greatly adds to the efficiency of the process.

This localized area of material removal is moved across the surface of the sheet thus generating a cut. Movement is achieved by manipulation of the focused laser spot (by CNC mirrors) or by mechanically moving the sheet on a CNC X-Y table. 'Hybrid' systems are also available where the material is moved in one axis and the laser spot moved in the other. Figure 1.44 shows schematic of laser cutting. The lens mount or the nozzle (or both) can be adjusted from left to right or into and out of the plane of the sketch. This allows for centralization of the focused beam with the nozzle. The vertical distance between the nozzle and the lens can also be adjusted.

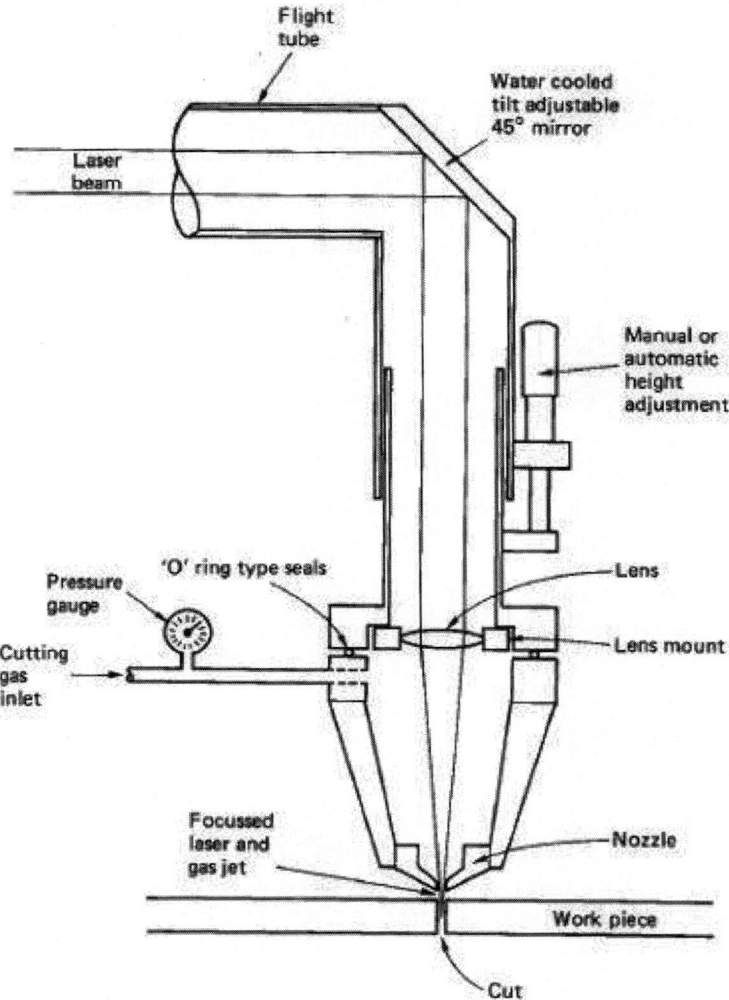

Figure 1.44   Schematic of laser cutting.

Laser cutting machine for textile

- Laser produces a beam of light which can be focused into a very small spot (0.25 mm), producing a very high energy density.

- The energy transfer to the material on which it focused, producing a rapid increase in temperature.

Cutting takes place by burning, melting and by vaporization. Figure 1.45 is laser cutting machine for textile.

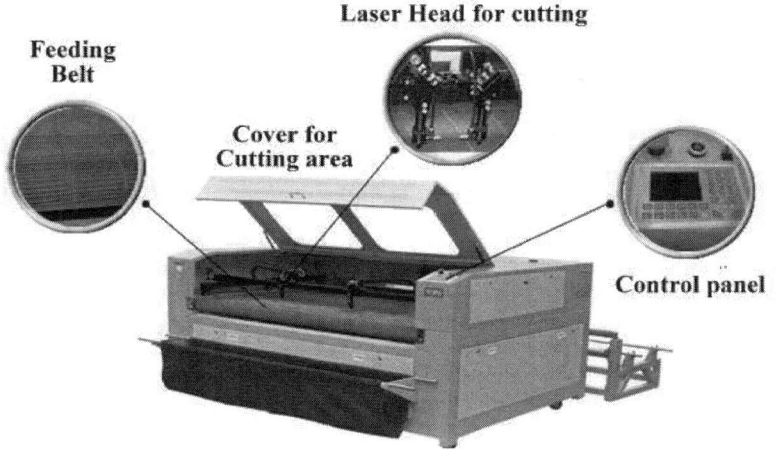

Figure 1.45 Laser cutting machine for textile.

Important components and conditions of laser cutting machine

Laser tube

The laser tube is fixed on the back side of the machine. It is one of the important components of the machine. The laser tube should be connected with the positive and negative pole. The electrode is the laser tube with spiral side as shown in Figure 1.46.

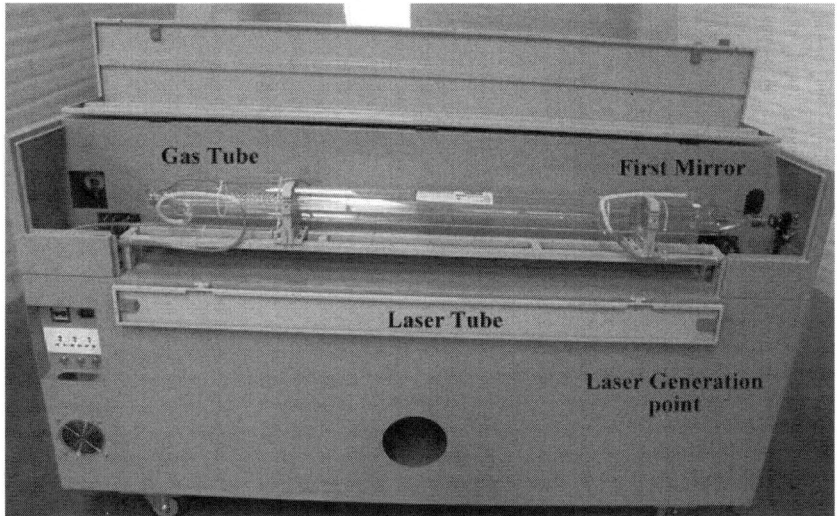

Figure 1.46 Laser tube set up in the laser cutting machine for textile.

Reflecting mirrors

During use of the machine, many reasons may occur for optical path offset, leading to the phenomenon of matt or light path is not correct. This may cause damage to laser tube and lead to serious problems. Hence the path of the laser beams should be maintained and checked always. The light from the laser tube can be launched in the first reflector centre. By adjusting the first mirror adjustment screws, light will hit the spot in the centre of the second mirror and by adjusting the third mirror in the third reflector attached to the optical block up, the optical head moves to the position closest to the second reflective mirror to make the light beam end and far-end completely overlap, and play spot light in the laser head into the centre of the hole (Figure 1.47).

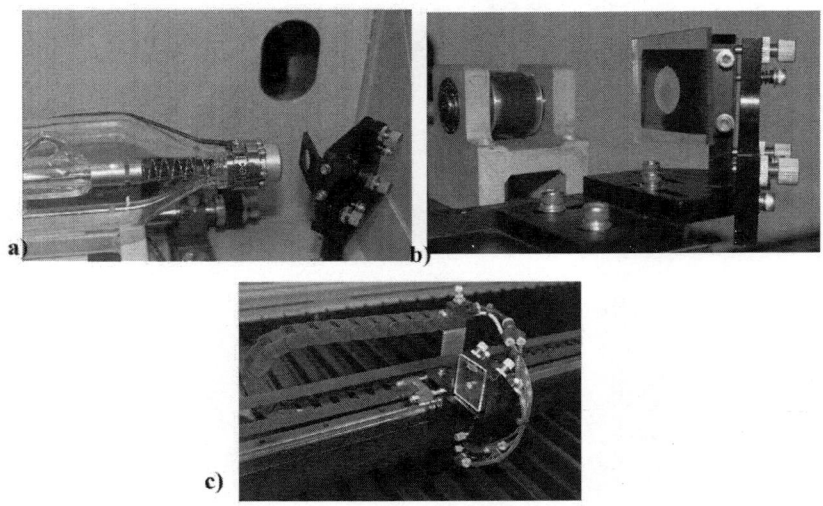

Figure 1.47 Reflecting mirrors in the laser cutting machine for textile (a) Mirror 1 (b) Mirror 2 and (c) mirror 3.

Cutting speed

The actual feed rate in use for a job will directly affect the cutting results; the feed rate is decidedly a function of the type of material and material thickness to be used. In any particular case, there will be some feed rate that is too high and the cut will simply fail to penetrate the material fully; at the other extreme, excessive heat input is likely to damage the material adjacent to the cut. In general, some feed rate closer to the maximum limit will be optimum, but always the choice is made experimentally on the basis of cutting results; the operator, with a little experience, can make this determination quite readily, making use of the feed rate override control.

Focal height of laser head

Focus assemblies provide support for the lens in order to image the beam. These assemblies generally provide means to adjust the focal point in or at the part. Height sensing devices can be incorporated to automatically maintain the proper focal point position regardless of undulations in the workpiece surface. These devices measure the lens-to-workpiece spacing either through contact probes riding on the workpiece surface or via a comparison of noncontact optical, acoustic, or electrical (inductance or capacitance measuring) signals bounced off the material. The feedback can trigger compensation of the vertical axis position.

Disadvantages

- The depth of focus is limited.

- Gives best results when the no of plies are less.

- Some time causes edge fusing.

- The main disincentives to the use of laser cutters are the quality of the cut edge (which may become charred and, with thermoplastics, may affect the feel of the edge), the possibility of less than 100% efficient cutting (as uncut threads can pull and adversely affect the visual appearance of the fabric), and the requirements to maintain the equipment.

- Lasers are not common for cutting garments, but they have been used successfully in home furnishings and in the cutting of sails (where edge fusing is actually desirable). They are growing in use for the production of patterns, labels and appliqués, often in conjunction with embroidery. There is a potential growth area with customized garments, particularly if the concept of mass customization becomes popular.

Advantages

- The process can be fully computer controlled. This, combined with the lack of necessity for complex jigging arrangements, means that, a change of job from cutting component 'A' out of natural textile to cutting component 'B' out of polymer/synthetic can be carried out in seconds.

- Although laser cutting is a thermal process, the actual area heated by the laser is very small and most of this heated material is removed during cutting. Thus, the thermal input to the bulk of the material is very low, heat affected zones are minimized and thermal distortion is generally avoided.

- It is a non-contact process, which means that material needs only to be lightly clamped or merely positioned under the beam. Flexible or flimsy materials can be cut with great precision and do not distort during cutting, as they would when cut by mechanical methods.

- Owing to the computerized nature of the process, the narrowness of the width and the lack of mechanical force on the sheet being cut, components can be arranged to 'nest' very close together. Hence, material waste can be reduced to a minimum. In some cases this principle can be extended until there is no waste material at all between similar edges of adjacent components.

- Although the capital cost of a laser-cutting machine is substantial, the running costs are generally low. Many industrial cases exist where a large installation has paid for itself in under a year.

1.4.3.5 Waterjet cutting machine

- A waterjet cutter is an industrial tool capable of cutting a wide variety of materials using a very high-pressure jet of water or a mixture of water. The cutter is commonly connected to a high-pressure water pump where the water is then ejected from the nozzle, cutting through the material by spraying it with the jet of high-speed water (Figure 1.48). The features of waterjet cutting machine is given below.

- The high-pressure jet acts as a solid tool when it encounters material in its way to cut, tearing the fibres on impact. Waterjet is most effective in harder sheet material, including leather and plastics. The water used must be filtered and deionized.

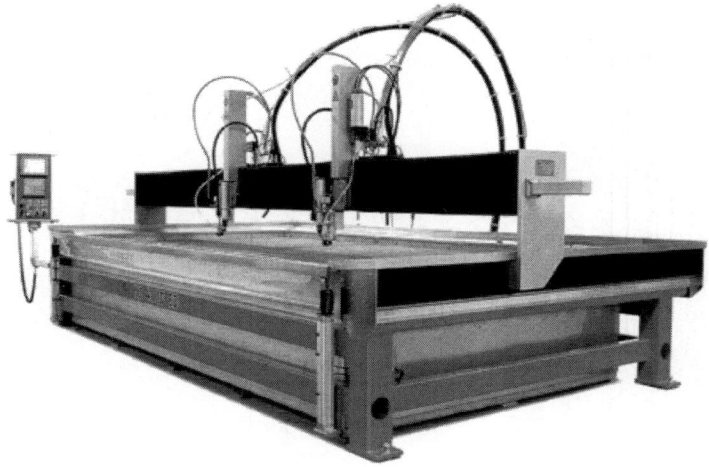

Figure 1.48   Water jet cutting machine for textile.

Features of waterjet cutting machine

1. In waterjet cutter, water or mixture of water is used to cut fabric which is controlled by computer.

2. Special software is used to cut fabric.

3. A very high velocity (60,000 lb/in.²) and small diameter steam of water is created by a waterjet intensifier pump.

4. The high pressure jet acts as a solid tool and sharp knife which can easily cut the fabrics, lathers and plastic materials.

5. As the jet penetrates successive plies in a spread, the momentum decreases and cutting ability is reduced. So, the lower lays of fabric cutting will be wide and less effective.

6. The jet of water and loose fibres normally caught and drains away by a catcher when cut the fabric. It is in the bottom of the fabric lays and moves with the same speed and the same direction of waterjet.

7. To improve the cutting speed, it is needed to adjust the pressure and radius of the jet.

Classification

The essential parameters of waterjet technology are the pressure and the water throughput. These two parameters produce the jet energy. The water is brought to a particular pressure with the pressure generator. The jet is generated in the cutting head, where the pressure energy is transformed into kinetic energy. Depending on the application, an abrasive is fed into the jet using a dosing system. The interaction between the jet and the workpiece produces the processing result.

Waterjet cutting (pure)

With pure waterjet cutting, a pure waterjet with a diameter of 0.1 mm cuts the material at up to three times the speed of sound (at speeds of up to 200 m/min). These materials include textiles, elastomers, fibres, thin plastics, food, paper, cardboard, leather, thermoplastic materials or food (Figure 1.49). The water is pressurized to 1000–6000 bar (standard approximately 3800 bar). After flowing through a high pressure needle valve, the water enters a 200 mm long and 3 mm in diameter wide collimation tube. It is then pressurized by a water nozzle or a dynamic pressure nozzle and accelerated. The jet speed varies according to geometry and pressure. The small diameter of the water nozzle produces a very high local energy density, which remains constant on a relatively long section in the direction of the waterjet and cuts cleanly and accurately when hitting the material.

Waterjet cutting with abrasives

With abrasive waterjet cutting, compact and hard materials such as metals (including steel), hard stone, glass (including bullet-proof glass) and ceramic are separated. Before the concentrated jet of water hits the material, a cutting material of the finest grain size (abrasive) is added in the required dose in a mixing chamber, which ensures micro cutting (Figure 1.49). The waterjet serves as an accelerator for the abrasive particles and hits the material with an impact speed of 800 m/s, thereby removing it with precision.

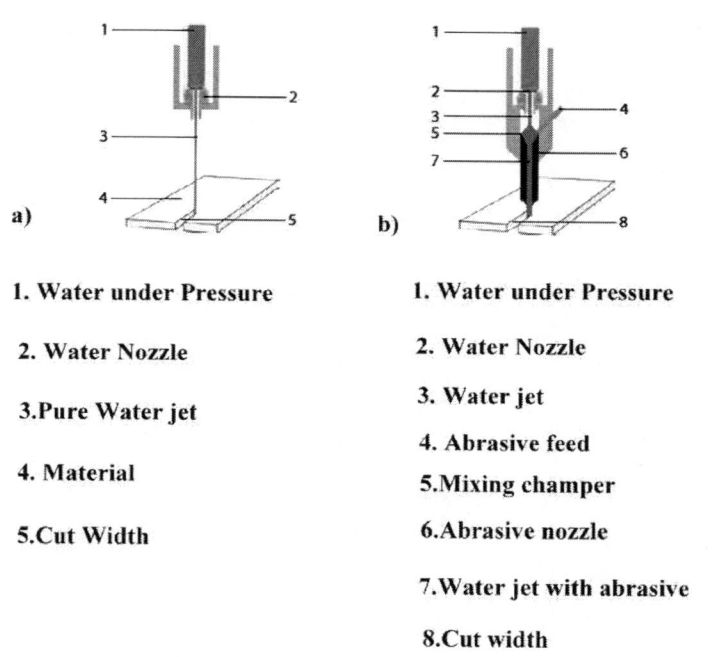

1. Water under Pressure

2. Water Nozzle

3.Pure Water jet

4. Material

5.Cut Width

1. Water under Pressure

2. Water Nozzle

3. Water jet

4. Abrasive feed

5.Mixing champer

6.Abrasive nozzle

7.Water jet with abrasive

8.Cut width

Figure 1.49   Working of (a) pure water jet cutting machine and (b) Water jet cutting machine with abrasives.

Until the waterjet is produced, abrasive waterjet cutting is identical to pure waterjet cutting. The difference is that the pure waterjet is no longer used just for cutting, but as a carrier material for the abrasive particles. The pure waterjet flows into a mixing chamber, into which the abrasive particles are then introduced. At the end of the mixing chamber is the focusing tube, in which the abrasive grains in the waterjet are accelerated and confined to a specific cross-section. After the focusing tube, the abrasive waterjet enters into the open and, after a few milli metres, hits the workpiece. The particles knock out all the crystals and also cut hard materials such as steel and glass.

Advantage of waterjet cutting machine

- Most effective to cut hard materials such as leather and plastic.
- Sound of cutting is less.
- Excess heat is not produce.
- Higher cutting speed.
- The table is not cut as catcher is used.
- Since there is no solid knife so, no sharpening is required.

Disadvantage of waterjet cutting machine

- There is a danger of wet edges.
- Water spot may occur on fabric.
- Hard water causes rusting. So water must be filtered and deionized before use.
- As in the lower lay, waterjet spreads out and then cut is wider and rough at the bottom of the spread.
- Not suitable for high lay of fabric. As the jet penetrates the successive plies in a spread, the momentum decreases and cutting ability reduces. It produces rougher cutting in bottom plies.
- The sound of jet need to control.
- Capital costs are high and tough for small companies to install this system.
- There is a danger of wet edges and water spotting.

1.4.3.6 Plasma cutting machine

Plasma is a thermally highly heated up, electrically conductive gas, which consists of positive and negative ions, electrons as well as of excited and neutral atoms and molecules. At the physics they often speak of the 4th state of aggregation.

As plasma gas, i.e. the monatomic argon and/or the bi-atomic gases such as hydrogen, nitrogen, oxygen and air are used. These plasma gases ionize and dissociate by the energy of the plasma arc. By recombination of the atoms and molecules outside the nozzle-cathode-system the received energy is suddenly set free and intensifies the thermal impact of the plasma beam on the workpiece. Figure 1.50 details the basic principle of plasma cutting process.

The plasma is additionally tied up by a water-cooled nozzle. With this energy densities up to 2×106 W/cm^2 inside of the plasma beam can be achieved. Because of the high temperature the plasma expands and flows with supersonic velocity speed to the workpiece (anode). Inside the plasma arc temperatures of 30,000°C can arise, that realize in connection with the high kinetic energy of the plasma beam and depending on the material thickness very high cutting speeds on all electrically conductive materials.

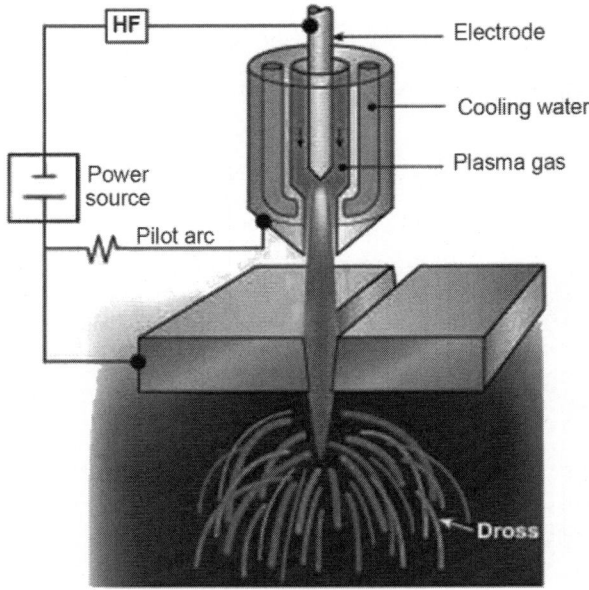

Figure 1.50   Basic principle of Plasma cutting process.

The term for advisable state of plasma arc is called stability of arc too. The stability of arc is keeping the plasma jet in desired form. It is possible to be provided by shape of plasma torch, streaming jet and water.

Other parameters are:

- temperature and electrical conducting,

- density of plasma jet,

- diameter of plasma beam,

- degree of the plasma beam focusing in output from nozzle.

Textile cutting by plasma

- Plasma is capable of cutting almost any man-made or natural material.

- However, plasma cutting is not commonly used for textiles.

- Its high temperature cutting method has the same disadvantages as those of laser cutting, causing synthetic materials to melt and form hard edges.

- Most of the plasma cutting machine used for metal and steel cutting purpose.

- In the area of textile the application of plasma cutter is still in research stage.

1.4.3.7 Ultrasonic cutting machine

An ultrasonic generator is used to control and provide energy to the ultrasonic head. A mechanical unit composed with a converter, a booster and a sonotrode/horn, to perform the cutting sealing operations.

Every ultrasonic unit contains the following five elements:

1. A POWER SUPPLY – which takes line power at 50 or 60 cycles and changes it to high ultrasonic frequency at 20,000 cycles/s or even higher

2. A CONVERTER – which contains piezoelectric crystals which change the incoming high frequency electrical signal to mechanical vibration.

3. A BOOSTER – which transmits the vibration energy and serves to increase its amplitude in much the same way as volume control on a radio.

4. A HORN – which delivers the vibration energy to the plastic film or fabric to be worked on.

5. AN ANVIL – or backup part which supports the workpiece and, in the case of textiles, takes the form of a pattern wheel or non-rotating cutter wheel depending on the application. The ultrasonic vibration is transmitted from the horn to the material developing frictional heat where they touch. This momentary heat fuses the edges of the fabric. If double plies are present, the plies join together. Where a cutting edge is used on the anvil, the fabric is cut through and the edges sealed at the same time.

The ultrasonic generator converts the power supply (100–250 V, 50–60 Hz) into a 20–30 kHz, 800–1000 V electrical signal. This signal is applied to piezoelectrical ceramics (included in the converter) that will convert this signal into mechanical oscillations. Figure 1.51 mentions the basic principle of ultrasonic cutting process. These oscillations will be amplified by the booster and converter, thus creating a hammer. The converter converts electricity into high frequency mechanical vibration. The active elements are usually piezoelectrics ceramics. The booster (optional) serves as an amplitude transformer. Amplitude magnification or reduction is achieved by certain design features or the geometrical shape of the booster.

The sonotrode, or horn, is the active part of the ultrasonic unit. It is in contact with the fabric and acts as a hammer against the counter-tool. This system will enable to melt the fibres within the fabric by mechanical action only (hammering), enabling to obtain a very clean and flexible melting without any burning or colour change. The sonotrode and counter-tool design will be adapted to the operation to be performed.

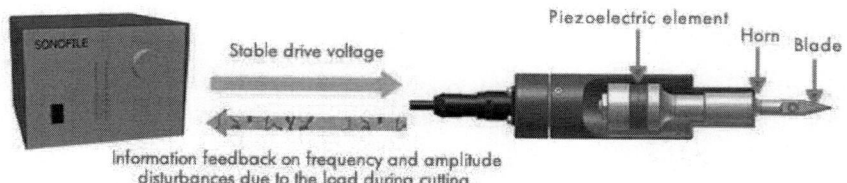

Figure 1.51   Basic principle of ultrasonic cutting process.

Applications

Slitting/sealing

Ultrasonic slitters cleanly cut and seal the edges of synthetic or blended fabrics, eliminating the disadvantages of hot wire or rotating knives. Fraying, unraveling, or beading along the cut edge is eliminated.

Loom cutter

Slitters may be mounted on a loom to cut the non-selvedge edge of woven fabric. Equipment is available to operate at the slow speed of weaving looms without burning or beading me fabrics. With special framework and motor drive, crosscuts can be made on the loom.

Hand cutting

A hand-held cutter is available for cutting straight lines or gentle curves by hand or using a simple guiding mechanism. One application is for a small flag and pennant manufacturer. Cutting the edge of the flags ultrasonically eliminates the former requirement of folding and stitching a seam along the edges, producing an attractive flag without excessive labour cost.

Plunge cutting

A plunge cut operator can be used to cut-to-length ribbons, belts, etc. Widths up to about 9 in. can be considered. A plunge approach is used together with a commercially available strip measuring device to preselect the length of ribbon and ultrasonically cut a diamond shape at the preselected length. The waistband of men's slacks is cut and seamed ultrasonically to give a pleasing finish to the cut edge without sewing.

1.4.3.8 End cutter

End cutter machines used in the cutting table to cut the fabric spread after a certain length (Figure 1.52). The end cutter may be manually operated or fully automated. In the case of manual operation, the cutter must be operated by the operator across the rail either by pulling or pushing the cutting head by a handle.

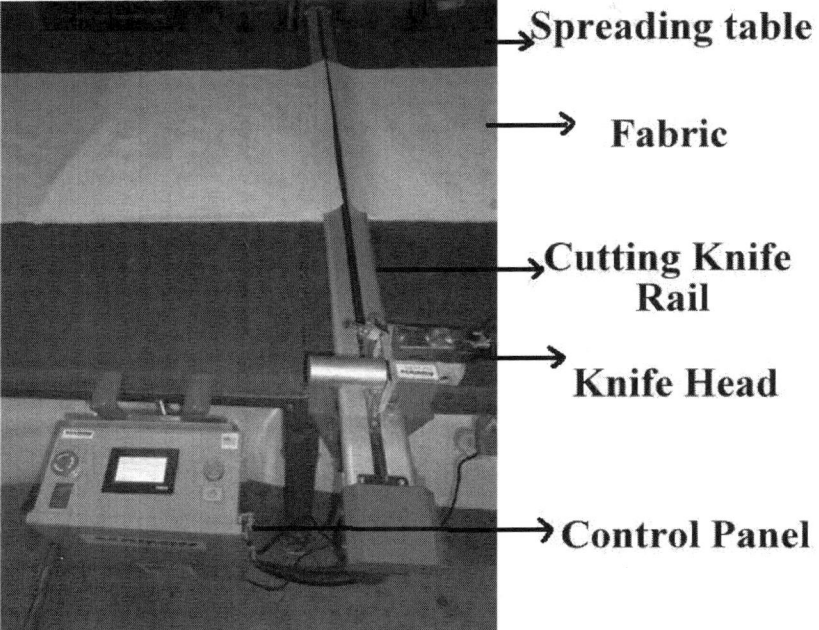

Spreading table

Fabric

Cutting Knife Rail

Knife Head

Control Panel

Figure 1.52   End cutter machine in textile fabric spreading application.

- The cutter and rail available for different width of the table from 1.22 m to 3.66 m of maximum.

- The manually operated machine comes with the standard 1 m handle to operate by the operator. However, the handle length can be modified based on requirement.

- In the case of automatic end cutter machine, the machine is connected with the spreading machine and the predefined length of the ply were fed in to the control panel by operator and hence after every spread completes the required length will be cut by the cutter.

- Automatic traversing of the cutting head with an automatic lift feature to enhance cycle time and throughput compared to manual models.

- As soon as the operator is finished pulling the ply down the table, the machine can be activated by remote control, which enables the cutting cycle to work.

- Compression pressure foot ensures a clean cut for both heavy and soft materials.

Maintenance procedure

Adjusting the blade sharpener: When the blade becomes worn down (smaller), adjust the grinder position by unscrewing the screw for the grinder arm to ensure a correct distance between the stone and the edge of the blade. After finishing the adjustment, re-tighten the screw.

Replacing the sharpening stone: Remove the grinding stone by simply unscrewing it. Mount the new grinding stone and re-tighten the screw.

Replacing the blade: Remove the sharpening stone by removing the mounting screws. Remove the guard from the front of the machine. Unscrew the lock nut for the blade. Remove the blade. When mounting the new blade, make sure that the side marked 'front side' is towards the operator. After the blade has been replaced, adjust the position of the sharpening stone as described above (Figure 1.53).

Figure 1.53   Round knife cutting machine assembly used in End cutter machine.

Replacing the lower blade: Remove the screw for the lower arm. Remove the lower blade arm and replace parts as needed. Replace the lower blade, ensuring the flat surface of the lower blade is adjacent to the blade edge.

Replacing the carbon brushes: The carbon brushes must be replaced when worn to 5–6 mm long. Too much wear will cause motor problems. Remove the carbon brush cap by turning it counterclockwise. Always replace both the right and left carbon brushes at the same time.

1.4.3.9 Strip/rib cutter

Strip or rib cutting machine (Figure 1.54) normally used in the apparel industry for the purpose of cutting 45° bias strip for the application of binding or piping. This machine also known as piping strip cutting machine or bias strip cutting machine.

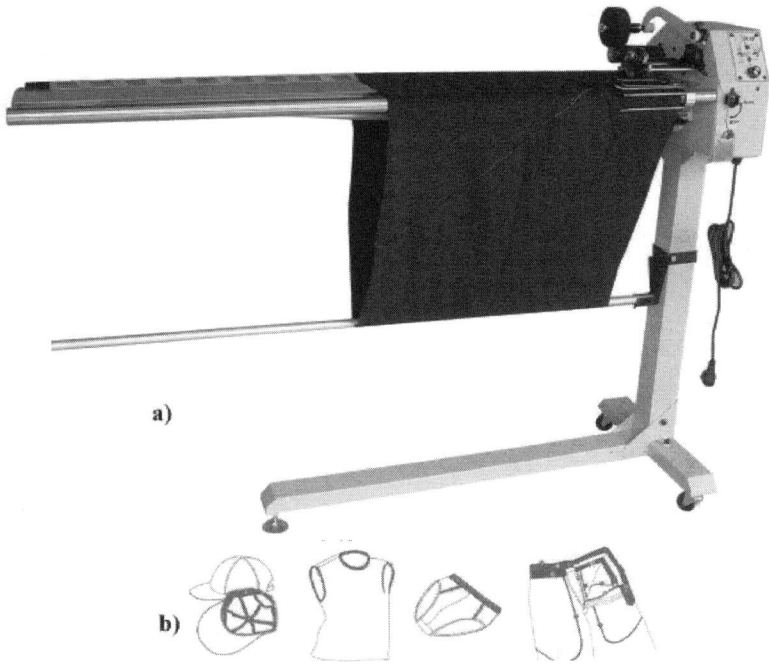

Figure 1.54 (a) Rib cutting machine for apparel industry and (b) application areas.

In this machine, the strip cut from continues cloth roll and wound as a strip roll. The application of the cut strip includes binding in T-shirts, trousers, vests, briefs, caps and leather accessories like shoes, etc. This various applications are shown in Figure 1.55.

- The main part of the strip cutter is the roller cutting knife and a winding roller. The cutting knife distance can be adjusted from the outer edge can be adjusted based on the strip width requirement.

- Measurement scale fixed before the knife aids and used to set up and monitor the width of the strip roll being cut.

- This machine can be used for all kind of material both knitted and woven.

- The maximum length of the fabric roll that can be used to cut as a strip is 110–135 cm.

- The maximum cloth circumference that can be used in the machine is from 60 to 200 cm.

- The machine running speed from 35 to 450 rpm and it will be able to complete a cloth roll in 45 s.

- The machine cut strip width varies from 5 to 90 mm based on the requirement.

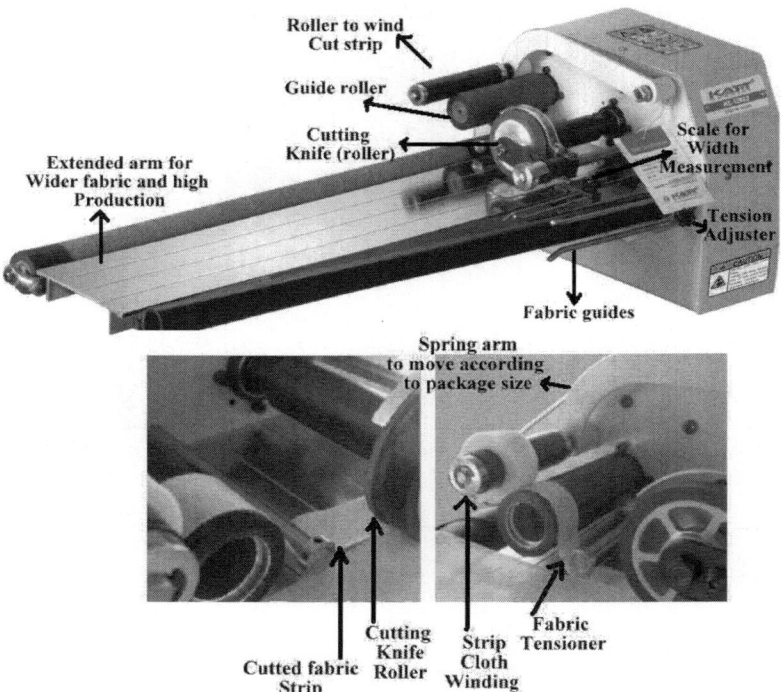

Figure 1.55   Different parts of rib cutting machine.

1.4.3.10 Rag cutter

- Rag cutting machine generally used in apparel industry as a means of making rags, disposable wipes and towels and also used in the at manufacturing facilities seeking solutions for repurposing expensive scrap materials (Figure 1.56). The machine works from a central motor, the drive from the motor is transferred to the cutting head by the belt mechanism. The cutting head fitted with the round knife and a sharpening device to perform the cutting operation.

- The cutter can cut the fabric clip, yarn hard waste, used garment, linen, plastic film, paper, nonwoven fabric, etc. with high efficiency. The size of the clips can be adjusted according to your requirement.

- The machine can accommodate two operators working on two workstations, both of which include an independent knife guard, sharpener and button removal trough.

- The motor is fully covered for the elimination of lint penetration.

- The weight of the machine is around 33 kg. The blade size is 15.2 cm and the cutting capacity 3.5–5 cm based on the manufacturer. The machine available in two or single cutting heads.

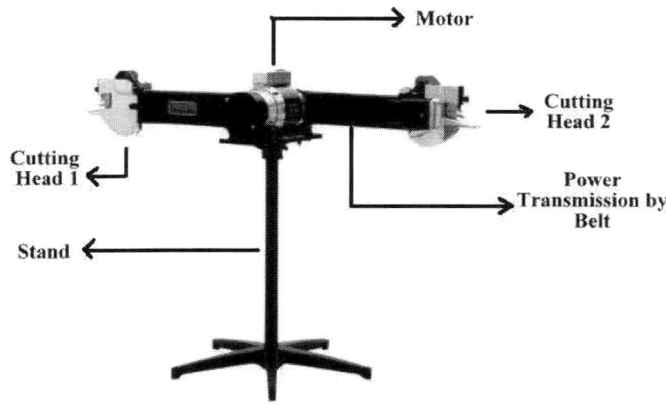

Figure 1.56 Rag cutting machine.

Features

- Equipped with a heavy-duty lubricated and sealed motor.

- Built-in sharpener and button remover.

- Telescopic height adjustment.

- Available in single or dual workstation models.

- Adaptable to many materials and applications.

1.4.4 Position markers used in cutting department

1.4.4.1 Notcher

- Many garment parts require that notches are cut in to the edges of them to enable alignment during sewing with other garment parts.

- The some of the cutting methods (band knife, round knife, straight knife, hand cutting) can be used to produce the notches, but accuracy depends upon the operator.

- Specialized notching equipment provides greater accuracy because a guide lines up the notcher with the cut edge to give consistent depth of notch at a consistent right angle to the edge.

There are two types of notches.

- Straight notches
- V notches

Further a hot notcher also available, which incorporated with a heating element in order that the blade may slightly scorch the fibres adjacent to the notch in order to prevent it fraying and disappearing. This cannot be used with thermoplastic fibres or certain unlined garments. The fabric requiring it may be loosely woven tweed. Figure 1.57 shows the notcher for fabric cut marking.

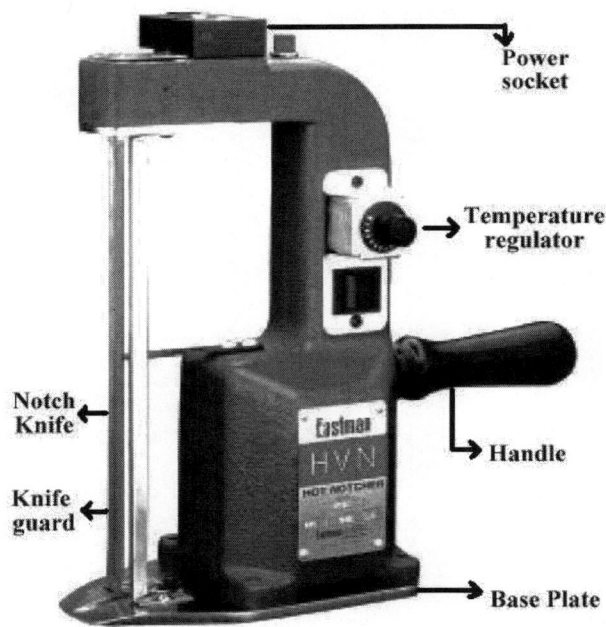

Figure 1.57   Notcher for fabric cut marking.

Types of notcher

1.   Cold Notcher – The cold notcher is a manually operated, spring-loaded device with a short blade mounted on a plunger. Placed at the edge of the bundle, the cutter lines the blade up with the notch. In a single

stroke downward, the notch is cut into the edge of all of the fabric plies.

2. Hot Notcher – When the fabric is a soft weave or knit, the cut notch will be lost in the edge fraying during handling each part. To create a more lasting notch, a hot notcher is used. The hot notcher utilizes a vertical heated edge which burns a notch into the edge of the bundle. The temperature is controlled, so as to leave a brown burn mark without melting or doing excessive damage to the fibres.

3. Ink Notcher – The ink notcher is similar to the hot notcher. Instead of burning a notch into the edge of the fabric, this device leaves a trace of UV marking ink on the edge of the fabric. This ink is visible under UV lights at the sewing station.

1.4.4.2 Cloth drills

Drill marks, which are round holes, are placed in cut components to show the ends of darts and the placement points of added components like patch pockets or flaps (Figure 1.58). These marks must be drilled so that after the darts are sewn, the marks remain in the material sewn into the dart. Normally, drill is used cold, hole remains visible until the sewing operator comes to use it.

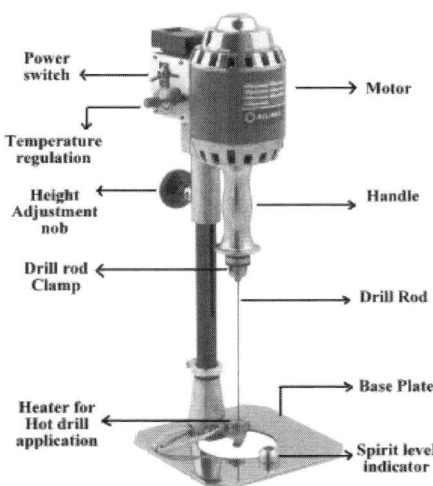

Power switch

Temperature regulation

Height Adjustment nob

Drill rod Clamp

Heater for Hot drill application

Motor

Handle

Drill Rod

Base Plate

Spirit level indicator

Figure 1.58   Drills for fabric cut marking.

When the marks are needed away from the edges of the garment, a hole is drilled through all the plies in the lay. For example, pocket position in centre of the body or dart position. The drills may be made in fabric spreads of up to 300 mm in height. The weight of the drilling machines varies between 5 and 13 kg. It consist of

- A motor
- Drill rod/needle
- Drill rod/needle guide
- Base plate with a hole
- A spirit level – to ensure the horizontalness of the base so that the drill will be vertical
- Operator handle

In some fabric, with looser weave, the hole may close up.

- In that case the hot drill is used, which will slightly scorch or fuse the edges of the hole.
- A hypodermic drill can also be used which leaves a small deposit of paint on each play of the fabric. But this is acceptable only where there is no problem of different colour deposition on the fabric.
- The dyes used in the drills are mostly water soluble and this method is preferably used in the garments which require washing after the production.
- It is important no mark remains in the fabric in all the cases of drill hole, fused edge or paint mark.
- All drill holes must be eventually concealed by the construction of the garment.
- A hole is drilled through the lay

Types of drill needles/rods

There are three different types of needles commonly available for drilling hole in the fabric. They are round point, half cup needle and hollow needle. These needles are available in different diameter as shown in Figures 1.59 and 1.60. It is advisable to drill the smallest hole possible, so as not to affect the finished garment.

Round point needle – used in closely woven material as it spreads the threads but does not cut them. In hot drill, the round point needle is always used with the heat for drilling loosely woven material.

The half cup needle – it cuts away the material and is used on coarser woven fabrics which would not show the marking of the round point needle.

The hollow needle – generally used for loosely woven fabric to cut into the hole in the material.

Diamond point: It is required when working with non-fusible, dense material such as natural wools or cottons. Bit is usually heated to separate and mark fabric.

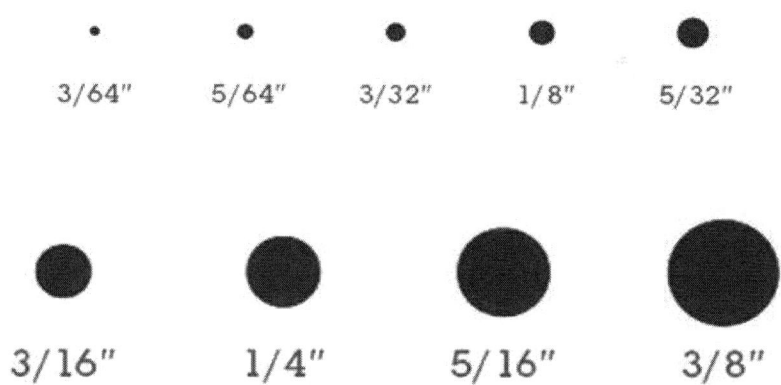

Figure 1.59 Different size of drill awls/rod for fabric marking.

Figure 1.60 Different shapes of drill awls/rod for fabric marking.

Taper point: It is required for fusible lightweight synthetic fabrics. Bit is usually heated to separate and mark fabric.

Open end awl: It is required when a definite hole is desired (material is removed rather than displaced). It is used primarily on dense materials for marking a wide range of materials.

Closed end awl: It is required for dense material when raveling is a problem.

Important machine settings

Needle clamp adjustment – Before drilling, the height of the clamp should be adjusted to limit the depth of the drilling. Without motor running, place the machine on the edge of the lay. Lower the motor bracket so that the lower tip of the needle enters the table (maximum of 0.8 mm), then lock the clamp height, so that the needle will not go lower when in use.

To operate hot drill – There will be a temperature regulation knob in the drilling machine, by turning it to the on position, the rod will be getting heat. Allow it for 5 min and once the rod heated, it can be used for drilling.

To change the needle – There is a chuck, hold the needle at the top of the holder. By opening the screw in it the needle can be removed. However, the needle guide has four different holes for different needle diameter. The needle must pass through the hole at the front of the needle guide to be in proper alignment.

During the needle change, rotate the guide, select proper size of the hole and use it accordingly.

1.4.4.3 Thread marker

- Thread marker is used in the cutting lays as a short term marking, whenever the long-term marking is not desired because of quality reasons. Figure 1.61 shows the thread marker.

- The machine uses a needle and a thread. The thread is pulled through the material in all the layers of the spread. When the components are cut into parts the thread goes with all the parts and which allows the operator to determine where to place pleats, darts, buttons or pockets.

- It can be also used for bundling cut pieces also in times, which ensures matched pairs by eliminating loose cut patterns of varying sizes.

- Maximum marking capacity is possible from 2 in. (5.08 mm) to 6 in. (15.2 mm) with simple looper mechanism for ease of operation.

- The width of the marking ranges from 1.19 mm to 1.98 mm.

- The machine weight around 5.75 kg.

1.4.4.4 Cut master

- Cut master (Figure 1.62) is a cutting machine generally used for the cutting applications where the small part like wide belting, hook and loop fastener, belt loop, nylon zipper, laces, webbings, etc. are needed to be cut for certain application for the apparel manufacturing purpose.

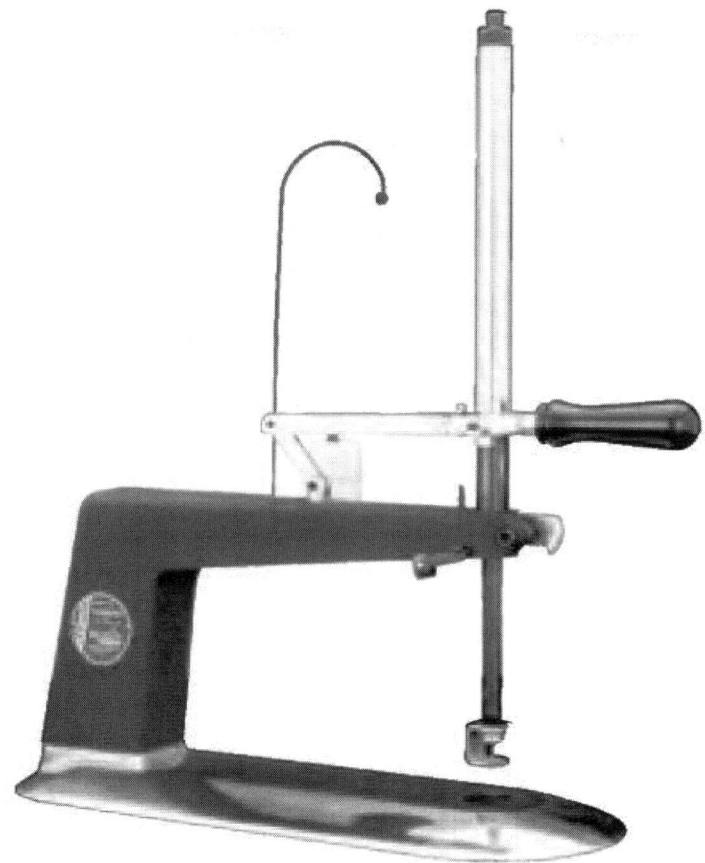

Figure 1.61 Thread marker.

- This cutting machine equipped with fully computerized control and a digital display programme for desired length and number of cuts.

- The machine consists of a roller holder, where the tape roll can be hold and then tension devices, to pass the tape with proper tension through a cutting knife with sensors. The machine automatically measured the length of the strap with sensors and also counts the total number of cuts as fed initially in the system

- The maximum cutting width of the machine is 10.16 cm (4 in.) and the maximum roll capacity is (15 in.). The machine weighs around 39 kg. The maximum cutting length up to 999.99 in. and the maximum number of piece count can be possible up to 999,999 numbers.

- The machine shut-off's automatically when the operation is complete as pre-programmed or the material supply depletes.

- The machine also comes with hot blade for cutting and sealing of the synthetic material woven material to avoid the unravel of the threads.

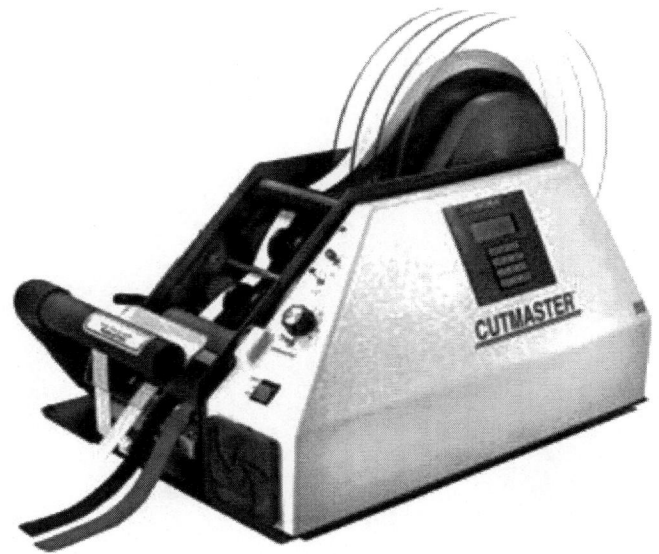

Figure 1.62   Cut master for strap / loop cutting purpose.

References

1. Solinger, Jacob, Apparel Manufacturing Handbook: Analysis, Principles and Practice, Columbia Boblin Media Corp., 1988.

2. Carr, H. and Latham, B., The Technology of Clothing Manufacture, 2nd edn. Blackwell Scientific, Oxford, 1994.

3. http://textileapex.blogspot.in.

4. www.gerbertechnology.com.

5. www.kuris.de.

6. www.eastmancuts.com.

7. http://www.maimin.com/.

8. http://www.cosmotex.net/spreading-machines.

9. http://exactagarment.com/.

10. http://www.shimaseiki.com/.

11. http://www.ttarp.com/.

12. http://www.sfcnclaser.com/.

13. Powell, J. and Kaplan, A., Laser cutting: from first principles to the state of the art, Proceedings of the 1st Pacific International Conference on Application of Lasers and Optics 2004.

14. Powell, J., The L I A Guide to Laser Cutting, Laser Institute of America. ISBN 0-912035-17X.

15. http://www.decoup.com/.

16. www.waterjet.ch.

17. www.waterjet.bystronic.com.

18. http://www.twi-global.com/.

19. Hatala, Michal, The Principle of Plasma Cutting Technology and Six Fold Plasma Cutting, 5th International Multidisciplinary Conference.

20. www.plasmaetch.com.

21. http://www.belsonic-machines.com/.

22. http://www.sonotec.com/en/tec.html.

23. www.focus-gmt-tech.com.

24. www.wolfmachine.com.

2

Sewing machine – mechanisms and settings

The important machines used in the sewing department of an apparel production industry like single needle lockstitch machine, overlock and flatlock machines are explained with their various parts and functionality. The different stitch formation mechanisms used in sewing machine along with their timing diagram are mentioned. The working mechanism of different sewing machines like single needle lockstitch machine (SNLS), double needle lockstitch machine (DNLS), overlock machine and flatlock machine are explained with clear illustration. The chapter also explains the important setting parameters that need to be checked in respective sewing machines in detail.

Key words: Sewing machine, Classification, Parts and functions, Stitch formation, Mechanism and setting points

2.1   Introduction

Sewing machines are textile machinery employed to stitch fabric or other material together with thread. Sewing machines were invented during the first industrial revolution to decrease the amount of manual sewing work performed in clothing companies. Since the invention of the first working sewing machine, generally considered to have been the work of Englishman Thomas Saint in 1790, the sewing machine has vastly improved the efficiency and productivity of fabric and clothing industries. Though some older machines use a chain stitch, the basic stitch of a modern sewing machine consists of two threads and is known as lockstitch. Industrial machines are usually specialized for a specific task, and so different machines may produce different types of stitch. Needle guards, safety devices to prevent accidental needle-stick injuries are often found on modern sewing machines. The most commonly used stitch in the olden times was chain stitch which had one major drawback of being very weak and the stitch can easily be pulled apart. When the machinery usage increased, it was understood that a stitch more suited to machine production was needed and hence usage of lockstitch emerged.

A lockstitch is created by two separate threads interlocking through the two layers of fabric, resulting in a sturdier stitch that looks the same from both sides of the fabric.

2.2   History and development of the sewing machine

In 1791 British inventor Thomas Saint patented a design for a sewing machine. The machine was expected to be used on canvas and leather. However, a working model was never built. In 1814, an Austrian tailor, Josef Madersperger, presented his first sewing machine for which he started the development in.

In 1830, a French tailor, Barthélemy Thimonnier obtained a patent for his sewing machine that sewed straight seams using chain stitch. He later went on to stitch sewing uniforms for the French army in 1841 with a capacity of 80 machines. His success was limited as French tailors who were afraid of losing their business started riots and destroyed his factory. Walter Hunt invented the lockstitch sewing machine in 1833. The machine developed by Hunt employed an eye-pointed needle (with the eye and the point on the same end) carrying the upper thread and a shuttle carrying the lower thread. The curved needle moved through the fabric horizontally, leaving the loop as it withdrew. The shuttle passed through the loop, interlocking the thread. The feed let the machine down requiring the machine to be stopped frequently and again set up. Hunt eventually lost interest in the machine and sold it without patenting it due to disinterest. In 1842, John Greenough patented the first sewing machine in the United States. Elias Howe patented his machine in 1845; using a similar method to Hunt's, except the fabric was held vertically. The major improvement he made was to put a groove in the needle running away from the point, starting from the eye. After spending a lot of time in England trying to attract interest in his machine he returned to America to find various people infringing his patent. He won his case in 1854 and hence got the right to claim royalties from the manufacturers using ideas covered by his patent. James Edward Allen Gibbs (1829–1902), a farmer from Raphine in Rockbridge County, Virginia patented the first chain-stitch single thread sewing machine on June 2, 1857. In partnership with James Wilcox, Gibbs became a principal in Wilcox & Gibbs Sewing Machine Company whose commercial machines are still used in the 21st century.

The original crochet stitch was introduced by Merrow and hence he won a case against Wilcox and Gibbs in the year 1905 for usage of the crochet stitch. The development of sewing machines was similar until 1900s when the first electric machines started to appear with more lavish decoration. Singer Sewing Co. developed the first electric sewing machine and introduced it in 1889. This machine had a motor strapped on the side. These machines became popular in households when more and more houses were electrified and slowly the motor was introduced into the casing. Modern machines may be computer controlled and use stepper motors or sequential cams which aid in achieving very complex patterns. As of now Asia has become the market leader in manufacturing modern machines and more specialized markets are starting to emerge.

2.3   Classification of sewing machine

2.3.1 Sewing machines classification based on their bed type

Sewing machines are mainly classified based on the bed type. Bed is the base of the sewing machine or it is actually the part on which the fabric rests while it is being sewed. The difference in bed types is attributed to the way the fabric moves with respect to the bed while being sewed.

The major bed type categories are the horizontal and vertical bed. The categories are based on the plane of the fabric sewing. The different types under each category are given in Figure 2.1. Some sewing machines are made in more than one bed types based on the production requirements.

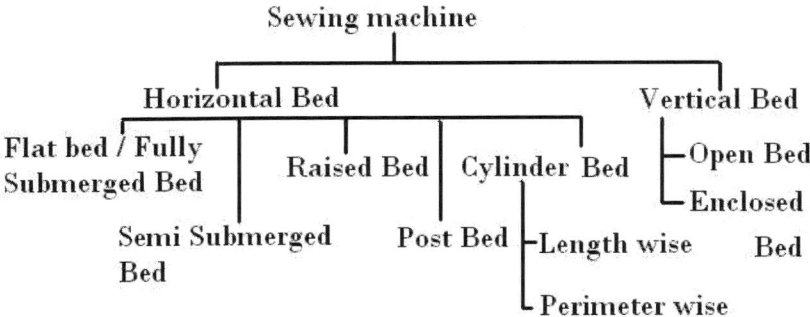

Figure 2.1   Sewing machine classification based on bed type.

2.3.2 Horizontal bed sewing machines

2.3.2.1 Flat bed/fully submerged bed

The name flat bed refers to the sewing machine frame which is built in a manner to enable the frame of the machine to be supported by the bottom side or the underside of the machine bed. In the flat bed type, the bed is mounted on a sewing machine table with the working surface of the bed coincides with the table top. When the machine table supports the bed arm of the machine, it is called as flat bed type (Figure 2.2).

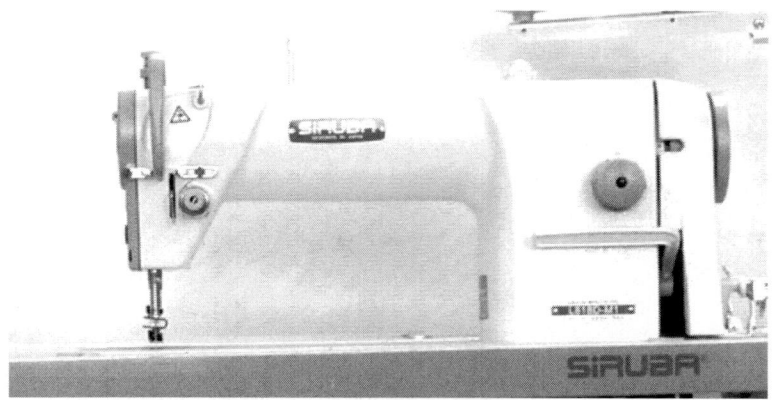

Figure 2.2   Flat bed Sewing machine.

2.3.2.2 Semi-submerged bed

Semi-submerged bed type has half of the sewing machine submerged below the table (mostly the oil tank is kept below the table) and the remaining part of the machine is kept above the table. Overlock machine is an example for semi-submerged machine type and normally the semi-submerged bed machines are used for sewing larger parts of the garments (Figure 2.3).

Figure 2.3   Semi-submerged sewing machine.

2.3.2.3 Raised bed

Raised bed type has the sewing machine frame supported by a frame section whose plane is below the bed arm section. In this type, the entire machine is mounted on the top of the table. It is used in the case where the requirement is for the garment to drape away from the needle for better handling (Figure 2.4).

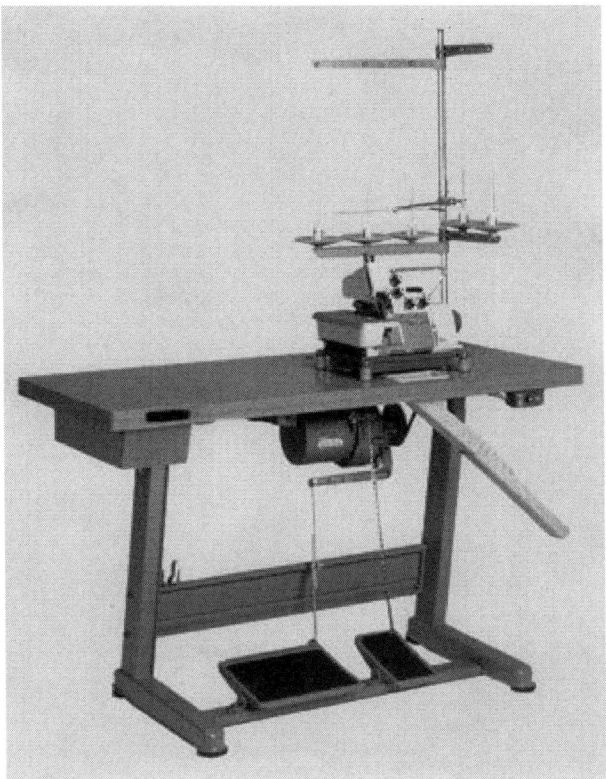

Figure 2.4   Raised bed Sewing machine.

2.3.2.4 Post bed

Post-bed sewing machine frame has the machine bed at the top of the post, and it sets vertically on the lower arm of the machine directly under the needles of the machine as shown in Figure 2.5. The complementary stitching mechanism is housed in the post when the machine has a feed system. Post bed machine which has the needle plate and foot mounted on a tall post enables sewing in tight spaces like inside sleeve cap, or shoes as the sewed part can follow its geometric inclination to cover this bed area as it is sewed.

Figure 2.5   Post bed sewing machine.

2.3.2.5 Cylinder bed

Cylinder bed is a sewing machine frame which permits one to sew a cylindrically shaped item in one of the following ways:

(i)   As the cylinder item is sewed, it travels into the bed part of the machine and encompasses the machine bed. In this situation, the sewed item must be extracted from the machine by retreating the item over the same path it travelled when it was sewed (Figure 2.6a). Such a cylinder bed is called *'Length Cylinder bed'*.

(ii) On the *perimeter cylinder bed*, the cylindrical item is sewed parallel to the circumference of the item as shown in Figure 2.6b. The perimeter of the items travels around the perimeter of the bed arm.

a)   b)

Figure 2.6   (a) Length Cylinder bed sewing machine, (b) Perimeter cylinder
               bed sewing machine.

2.3.2.6 Off – the – arm machine

The frame has a tube like section for the bed of the machine. This tubular
bed may be exactly the same size as the bed arm of a length cylinder bed.
The fabric travels on to the bed arm of a length cylinder bed. Such machine
permits an operator to sew cylinder, such as sleeves with lapped type seams,
the machine is shown in Figure 2.7.

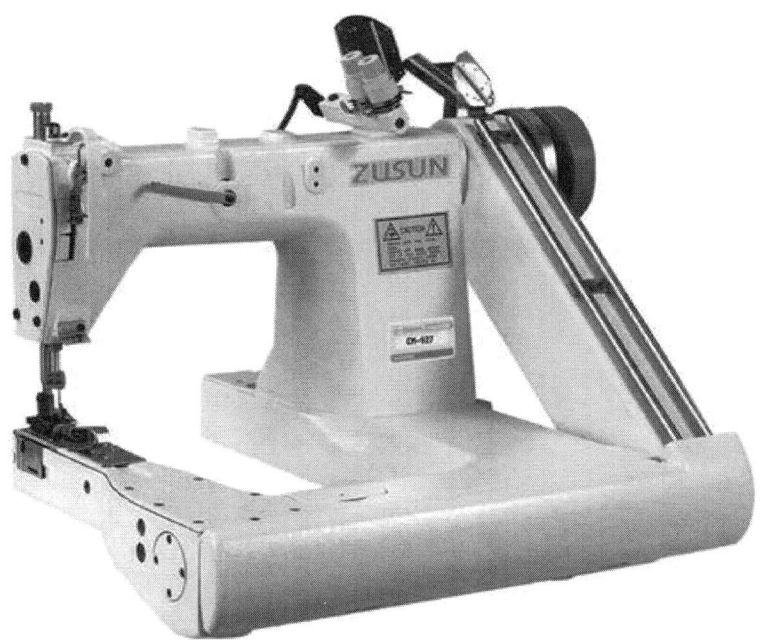

Figure 2.7   Feed of f the arm sewing machine.

2.3.3 Vertical bed machine

2.3.3.1 Open vertical bed

It is a frame in which the fabric is suspended vertically from the machine it sewed, without being enclosed by the machine frame. The supporting plane of a vertical frame is parallel to the horizontal.

2.3.3.2 Closed vertical bed

In the closed vertical bed the suspended fabric is surrounded by sections of the frame while the fabric is sewed. Such a frame limits the fabric size that can be sewed with ease.

2.3.4 Sewing machine – classification based on operator control

In general the machines are classified as follows with respect to operator interface.

2.3.4.1 Manual control

The operator controls manually all the phases of the sewing cycle. Initially from the picking up of the fabric, positioning the fabric in the sewing machine, all sewing and re-positioning of the fabrics during sewing, extracting and discarding the fabric after sewing.

2.3.4.2 Semi-automatic sewing machine

In these machines, sewing is controlled automatically by the machine after the operator actuates the machine such as buttonhole machines and bar tack machines. These machines are called stop motion machines. There are two sub classes in semi-automatic machines.

(i) Operator picks the part to be sewed, positions initially, extracts and moves manually

(ii) Operator picks up the part to be sewed, positions initially but the machine extracts and moves automatically after the sewing phase completed

2.3.4.3 Automatic sewing machine

In this machine the operator merely loads a hopper in the machine with the stack of garment parts. The machine automatically picks up, positions, sews, extracts and moves the sewed pieces.

2.4   Single needle lockstitch machine

2.4.1 Parts and functions

The single needle lockstitch machine is one of the most versatile machines used in all kinds of applications and irrespective of the fabric type used, either knitted or woven. This machine is manufactured by different manufacturers; Figure 2.8 represents the most important parts of single needle lockstitch machine.

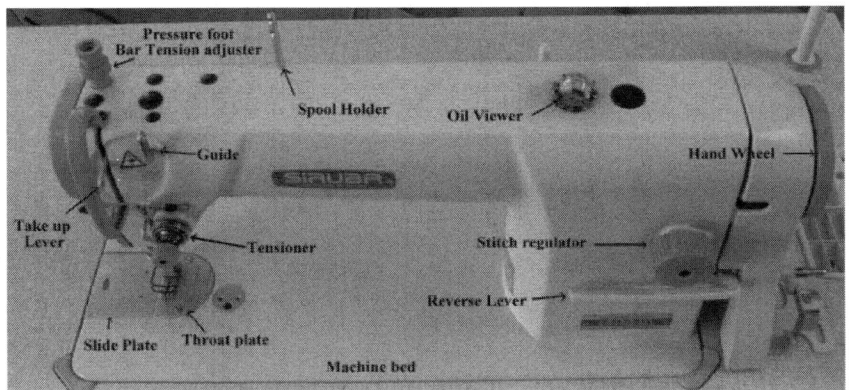

Figure 2.8   Single needle lock stitch sewing machine.

2.4.1.1 Spool holder

The sewing machine spool holder does more than just holding a spool of thread. The spool holder is the first step for evenly feeding the thread to the sewing machine needle. There are many spools of thread and many spool holders to achieve this vital first step the thread will take as it travels to the sewing machine needle (Figure 2.9).

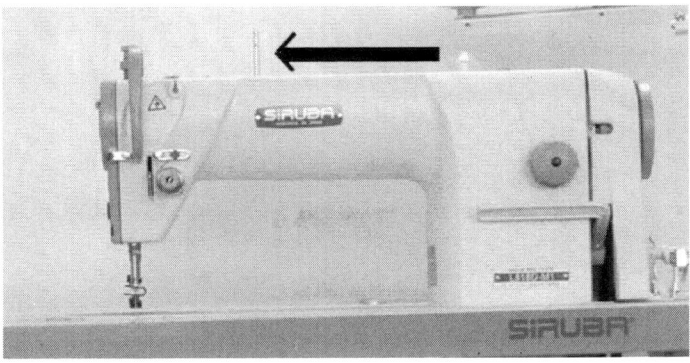

Figure 2.9   Spool holder in a sewing machine.

2.4.1.2 Thread cutter

Sewing machine thread cutters are usually located behind the needle area of the sewing machine, so that it is convenient as the fabric is removed at the back of the machine, the thread can be cut with the thread cutter. The sewing machine thread cutter does not cut the thread close to the fabric. The thread close to the fabric has to be trimmed with scissors.

2.4.1.3 Bobbin case

The sewing machine bobbin case may be removable from the sewing machine or it may be built in the sewing machine. The main function of the bobbin case is to hold the bobbin, which carries the looper thread for single needle lockstitch machine. Bobbin cases are not interchangeable in different sewing machines. All bobbin cases have a threading pattern. A bobbin is placed into a removable bobbin case so the thread and the slot form an upside-down 'V'. The important parts of the bobbin case are given in Figure 2.10. The thread will be passed through the slot and tension spring and delivered through the delivery eye. The tension spring controls the bobbin thread tension. By adjusting the tension spring screw, we can modify the bobbin thread tension and hence we can control the thread flow from the bobbin.

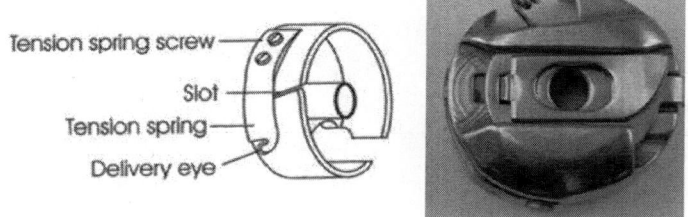

Figure 2.10   Bobbin case for single needle lock stitch machine.

2.4.1.4 Bobbin

The bobbin provides the thread for the underside of the stitches a sewing machine forms. Bobbins are not interchangeable between sewing machines (Figure 2.11). Bobbins are filled using the bobbin winder and the thread should be evenly distributed on the bobbin. It is advisable to have extra bobbins for the sewing machine so that unwinding of bobbins to change the colour of the thread on the sewing machine can be avoided. The bobbin spool holds approximately 70 yards of thread. The exact amount of thread held by a bobbin depends on the relative size of the bobbin thread used. Bobbins are enclosed in bobbin cases around which the bobbin hook rotates as the machine operates. The bobbin hook grasps the thread at the scarf point and carries the loop of the needle thread around the bobbin case which holds the bobbin and its thread.

Figure 2.11 Bobbin for single needle lock stitch machine.

2.4.1.5 Bobbin winder

Bobbin winding mechanisms are a bit different for every sewing machine. Bobbin winders can be found on the top, front or end of the sewing machine. Some sewing machines have a bobbin that is filled in place in the bobbin case. Most of these machines have a lever to lock the machine into winding the bobbin without un-threading the sewing machine. Most industrial sewing machines have a bobbin winder that is separate from the machine, usually located on the table of the machine; for more details of the winding mechanism see Figure 2.12.

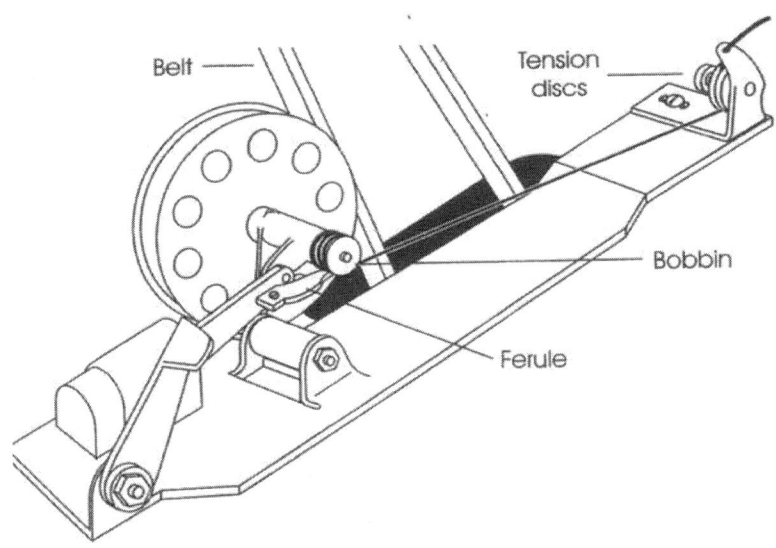

Figure 2.12 Bobbin Winder.

2.4.1.5.1 Parts of a bobbin winder

The thread starts on the thread spool holder. Thread guides are a vital part of the bobbin winding process to maintain tension on the thread and achieve an evenly wound bobbin. The bobbin holder is movable in most machines. The bobbin must be placed all the way into place on the bobbin holder. The bobbin holder is away from the brake when the bobbin is being put on and taken off of the holder. It slides towards the brake for the bobbin winding process.

The brake stops the bobbin or changes the sound of the machine so you know when the bobbin is full. It is not advisable to overfill a bobbin.

2.4.1.5.2 Basic threading process for bobbin winding

The spool of thread is placed on the spool holder. Thread guides feed the thread to the bobbin with an even tension. The bobbin is placed on the bobbin holder and thread is hand wound a few times around the bobbin or fed through a hole on the bobbin. Then the bobbin holder is slid with the bobbin towards the bobbin brake. Foot pedal is used to activate the machine with slow even speed.

When the brake stops the bobbin from turning or when a change in the sound of the machine motor is heard, then bobbin winding should be stopped. The bobbin holder should be slid away from the brake and the thread needs to be cut before removing the bobbin.

2.4.1.5.3 Disengaging the machine while winding a bobbin

Many sewing machines require an adjustment by turning the inner portion of the hand wheel to disengage the machine needle area while winding a bobbin.

2.4.1.6 Tension disc

Tensions discs control the amount of pressure applied to the thread for an even feed to the machine needle causing an evenly formed stitch. Many newer sewing machines have the discs hidden inside the machine casing. On machines with hidden tension discs there is usually electronics controlling the tension on the discs. The presser foot should be always raised when threading the sewing machine, so that the discs are not engaged and are open to accept the thread.

Adjusting machine tension

Sewing machine tension is adjusted when the bobbin and upper thread is not even on both sides of the fabric. The intersection of the bobbin and the upper thread should be between the two layers of fabric when sewing is done for two layers of fabric.

The main functions of tension device are

1. It positions the thread to needle

2. It regulates the flow of the thread

3. It maintains the smoothness in stitching

4. It controls the thread passage precisely

There are 2 types of tension devices. Both the devices contain the following components: two pressure discs, tension spring, thumb nut pressure control, the tension mounting bar and the pressure releasing units as shown in Figure 2.13.

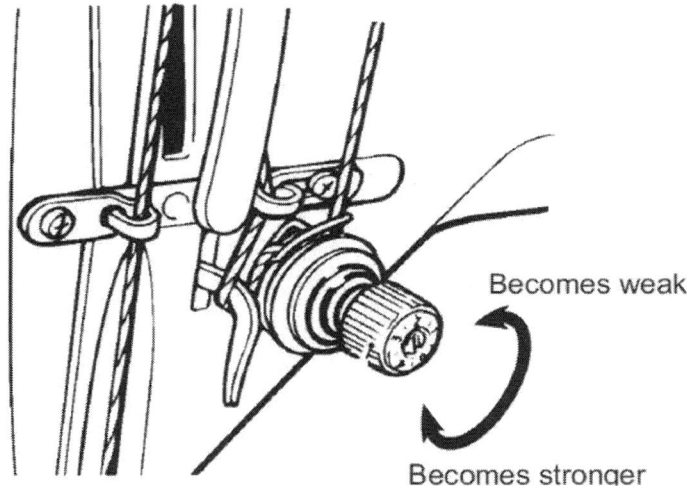

Becomes weak

Becomes stronger

Figure 2.13   Tension device.

(a)   Direct tension device

In direct link system, the two pressure dials, the tension spring and the thumb nut are mounted in that order on the post which is in turn inserted into the machine head and held in a place by stud screw.

(b)   Indirect tension device

The indirect device has another component called tension wheel.

The thread must be guided between the pressure discs in a manner which will keep the thread passing against the tension bar constantly as the thread consumed by the machine. The tension bar is kept in an angle less than 180°, to avoid the thread passage changes during the machine running. The pressure disc extends constant pressure on the thread as it rides against the pressure bar. The thumb nut controls the compression of the tension spring, which in turn presses the tension disc.

2.4.1.7 Stitch length regulator

The stitch length adjustment adjusts the length of the stitches made by the sewing machine. The adjustment takes place at the feed dog and not at the machine needle. Shortening the stitch length shortens the amount of fabric that is fed under the presser foot before the needle comes down. Lengthening the stitch length lengthens the amount of fabric that is fed under the presser foot before the needle comes down (Figure 2.8).

2.4.1.8 Stitch width adjustment

Stitch width adjustments allow to vary the width of the stitch. This adjustment became available with zigzag capabilities being built into sewing machines. The maximum width available varies with the sewing machine.

2.4.1.9 Presser foot

The presser foot exerts downward pressure on the fabric as it is fed under the needle. Lowering the presser foot engages the tension discs. The presser foot is used for all regular straight sewing. If the machine has zigzag capability, this foot has an opening wide enough for the needle in any position (Figure 2.14). The major functions of the presser foots are given below.

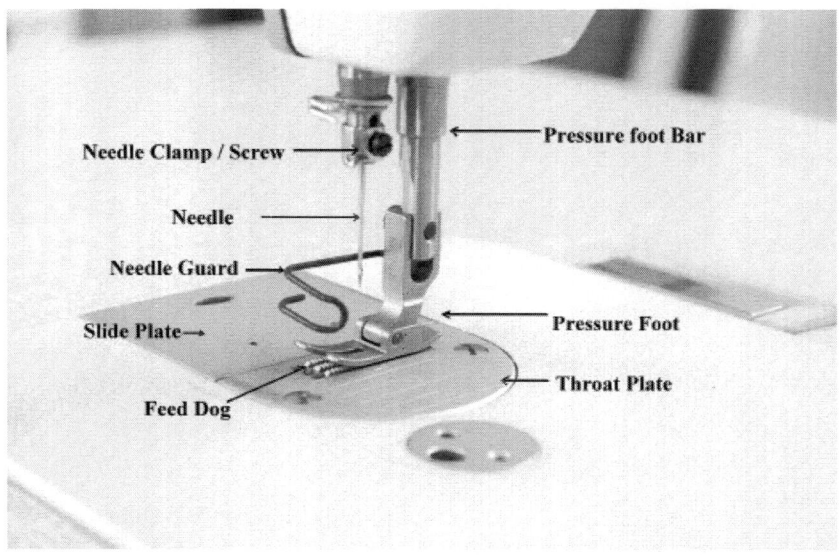

Figure 2.14　Parts around the feeding point of sewing machine.

- It stabilizes materials to sew jointly on the surface of throat plate, and determines the sewing position.

- It presses the materials so that materials are not lifted with the needle when needle comes out of materials.

- It makes materials come in close contact with teeth of feed dog with adequate pressure, so that the sewing direction is not disturbed when feed dog feeds materials forward or backward.

2.4.1.10 Feed dog

The feed dog feeds the fabric under the presser foot while the fabric is guided. The feed dog regulates the stitch length based on the amount of fabric passes under the presser foot as the machine stitches as shown in Figure 2.14. It is important to not push and pull the fabric under the presser foot. The feed dog should be allowed to move the fabric, so that sewing machine needles are not bent and broken. Most feed dogs are metals but there was a period of time that they were made in rubber. The rubber feed dogs wear out easily and had to be replaced if they had rounded edges or showed signs of wear.

2.4.1.11 Slide plate

The sewing machine slide plate provides access to the bobbin area of a sewing machine and protects the bobbin area from thread and fabric being caught in moving parts when the machine is operating. Most slide plates do slide out of the way to access the bobbin but some have a button or lever to release the slide plate. It is shown in Figure 2.14.

2.4.1.12 Throat plate

The throat plate is a removable part, which protects the bobbin and underside of the sewing machine. It is attached by screws to the bed arm of the machine. The needle passes through the needle hole in throat plate during the sewing cycle. The throat plate has seam guides (Figure 2.14). It has to be verified that the machine seam guide is accurate with the needle position being used. Many newer machines have built in visibility of the bobbin area to ensure the bobbin is filled before starting a sewing work.

2.4.1.13 Take-up lever

The take-up lever is an important part in the sewing machine for threading and for knowing the upward position of the sewing machine needle (Figure 2.15). The take-up lever is always kept at the top when placing or removing fabric from under the presser foot to prevent snagging and bending the needle. The machine is threaded with the thread in the take-up lever to prevent knotted wads of thread. The take-up-lever pulls the thread back after each stitch, so that the stitches will lay evenly into the fabric.

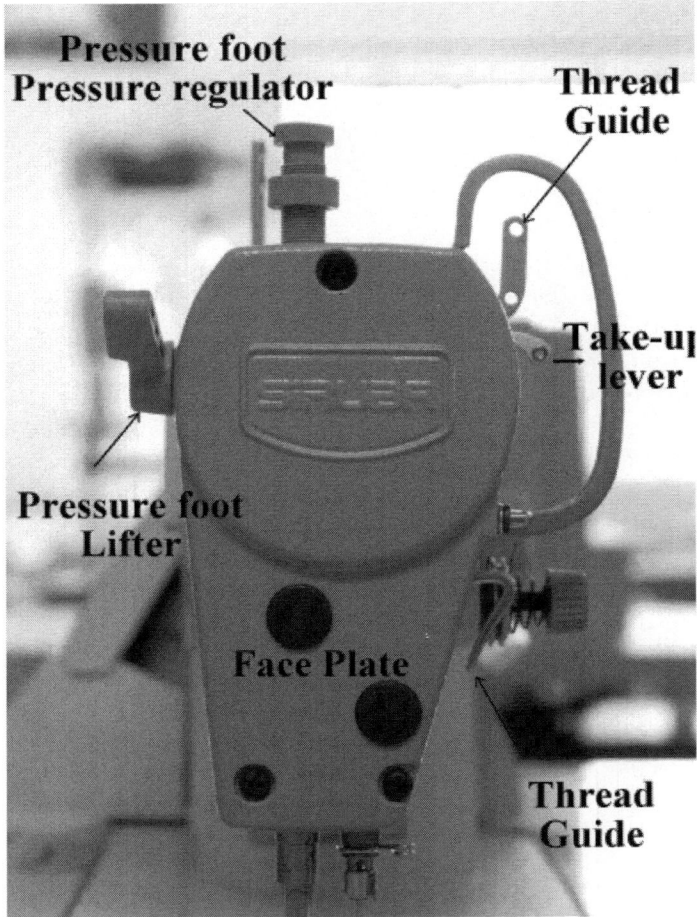

Figure 2.15   Side view of the sewing machine.

2.4.1.14 Pressure adjustment for presser foot

The pressure adjustment sets that amount of pressure that the presser foot will exert on the fabric. Downward pressure on the presser foot holds the fabric between the feed dog and sewing machine foot. The amount of pressure would need to vary when switching between lightweight sheer fabric and multiple layers of denim for the machine to feed the fabric.

Pressure adjustments

The pressure dial determines how much pressure the presser foot will have on the fabric. If the fabric slips when the presser foot is down, then the pressure on the presser foot should be increased. If the fabric is held so tightly that the feed dog is unable to move the fabric, decrease the amount of pressure on the presser foot (Figure 2.15).

2.4.1.15 Presser foot lever

The presser foot lever raises and lowers the presser foot. The amount of pressure exerted on the presser foot is controlled by the pressure adjustment. The presser foot lifter gently lifts the presser foot up and lowers it against the feed dog (Figure 2.15). When the presser foot lever is in the upward position, the tension discs are disengaged. When the presser foot lever is down the tension discs are engaged. The presser foot must be lowered before beginning to sew as fabric will not move through the machine if it is not lowered.

2.4.1.16 Face plate

The face plate is a machine side cover as shown in Figure 2.15; it covers the internal mechanism of needle and presser foot. By opening the face plate, the presser foot tension mechanism and the needle bar height can be adjusted. This face plate also covers the driving forces for the take up lever and presser foot lifter mechanisms inside it.

2.4.1.17 Foot pedal

The sewing machine foot pedal controls the speed of the sewing machine based on the pressure applied to the pedal similar to a car gas pedal working. Sewing machine foot pedals are usually placed on the floor as shown in Figure 2.16, but in some cases they are located in a cabinet with a knee lever that applies pressure to the foot pedal.

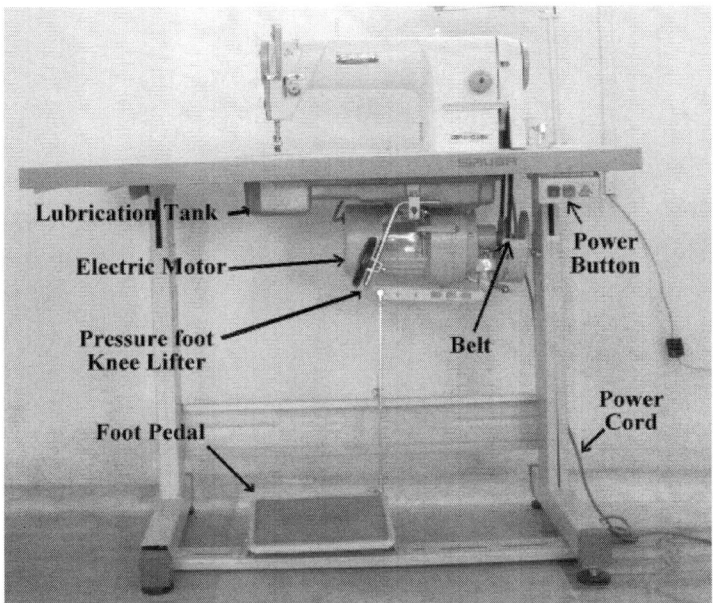

Figure 2.16   sewing machine motor.

2.4.1.18 Power cord

The power cord feeds electricity to the machine. The power cord should always be tightly connected to the sewing machine for a consistent flow of electricity (Figure 2.16).

2.4.1.19 Hand wheel

The main purpose of a hand wheel on a sewing machine is to slowly turn the needle by hand. This gives you control to position fabric under the needle and align fabric to guides built into the machine as in Figure 2.16. On older model sewing machines the hand wheel has an inner wheel that is loosened to allow the machine to wind a bobbin while disengaging the needle area of the sewing machine. This is done by holding the outer wheel and turning the inner wheel. Newer electronic machines have a simple hand wheel that does not have a disengaging feature. This feature is built into the machine when you move the bobbin towards the bobbin winding brake.

2.4.1.20 Motor

Sewing machine motors have evolved to being enclosed inside the sewing machine case. External motors are usual only in the case of industrial sewing machines, where the motor is located below the sewing machine table as shown in Figure 2.16.

2.4.1.21 Sewing machine belts

Many older sewing machines had visible belts. Some sewing machines have belts that are hidden inside the case. Straight belts are used for sewing machines (Figure 2.16).

2.4.2 Sewing needles

The function of the sewing machine needle is general are,

- **To produce a hole** in the material for the thread to pass through and to do so without causing any damage to the material.

- **To carry** the needle **thread** through the material and then **form a loop** which can be picked up by the hook on the bobbin case in a lockstitch machine or by the looper or other mechanism available in the machine.

- **To pass** the needle thread through the loop formed by the looper mechanism on the machines other than lockstitch.

The important parts of a normal sewing machine needle are given in Figure 2.17 and the name of the parts are: A - Shank diameter, B - Shoulder diameter

(B1 - Needle size, B2 - Trunk), E - Length of needle, D - Butt to eyelet, J - Length of eyelet, K - Width of eyelet, N - Length of shank, H - Length of scarf, Z - Depth of groove, F - Length of point.

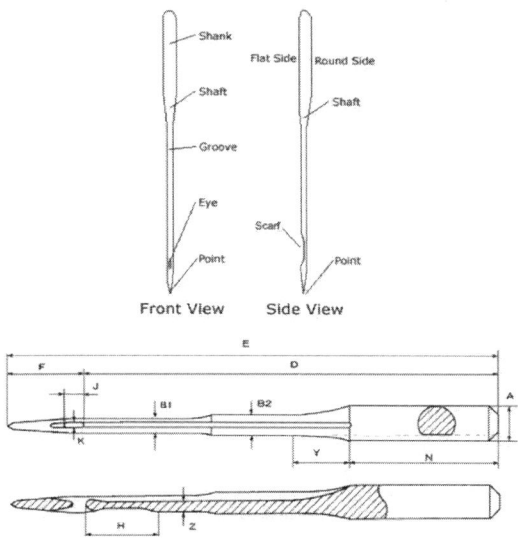

Figure 2.17   Sewing needle.

Shank diameter (A): It mainly divided into the following three systems. For overlock machine needle it is 2.02 mm and for lockstitch system, it is 1.62 mm–1.90 mm. For special sewing machine system such as straight buttonholing it is 2.00 mm.

Butt to eyelet (length between top end of eyelet and top end of shank) (D):

This is the most important dimension for hook or looper to scoop thread loop, and the length of D is fixed even when thickness of needle (needle size no.) varies.

Length of needle (E):

For DB × 1, DP × 5, etc., whenever the shank gets thicker, the length of needle gets longer. DC type needles are for overlock and chain stitch and the total length is fixed since it is required to scoop looper thread at the needle tip.

Thickness (needle size)

Thickness is dimension B1 and shown as needle size. Generally, a needle consists of 2-step stretched wire in which there is the trunk B2 thicker than the trunk B1. (DC × 1 and DC × 27 consist of one-step stretched wire since the whole length is short.) Dimension B1 (needle size) prevents the needle from vibration and protects the rise of needle heat by reducing friction when the needle comes off cloth.

Length of shank (N):

If the length of shank N gets longer, it is better for needle-wobbling or needle-bent. However, if the shank portion enters material, it will cause material breakage or puckering. As a result, the length within the range that the shank does not enter material is good.

Shape of scarf

The typical shapes of scarf are of standard type and of boat type. The boat type shape is good for making needle thread loop and effective to protect stitch skipping. However, the blade point to scoop needle thread should be positioned at the height where it does not come in contact with the lower portion of scarf. In addition, resistance at upper and lower angle portions of the scarf slightly increases when raising or lowering material.

2.4.2.1 Sewing needle parts and functions

Shank

The shank is the upper part of the needle which is located within the needle bar. It may be cylindrical or have a flat side, based on how it is secured into the machine. It is the support of the needle as a whole and is usually larger in diameter than the rest of the needle for better strength.

Shoulder

The shoulder is the section which is intermediate between the shank and the *blade*, the latter forming the longest part of the needle down to the eye. The blade is subject to the greatest amount of friction from the material through which the needle passes. In needles designed for use in high-speed sewing machines, the shoulder is often extended into the upper part of the blade to give a thicker cross-section which just enters the material when the needle is at its lowest point on each stitch. This supplementary shank or reinforced blade strengthens the needle and also enlarges the hole in the material when the needle is at its lowest point, thus reducing friction between it and the material during withdrawal after each stitch. Alternatively, the blade can be gradually tapered along its length from shank to tip, as another way of reducing friction. However, this friction can create enough heat to cause serious problems when sewing synthetic fibres.

Long groove

The long groove in the blade provides a protective channel in which the thread is drawn down through the material during stitch formation. Sewing thread can suffer considerably from abrasion during sewing as a result of friction against the fabric and a correctly shaped long groove with a depth matched to the thread diameter offering considerable protection to the thread.

Short groove

The short groove is on the side of the needle which is towards the hook or looper; it extends a little above and below the eye. It assists in the formation of the loop in the needle thread.

Needle eye

The eye of the needle is the hole extending through the blade from the long groove on one side to the short groove on the other. The shape of the inside of the eye at the top is critical both in reducing thread damage as the needle penetrates the material and in producing a good loop formation. On some needles, known as bulged eye needles, the eye area has a larger cross-section than the rest of the blade. This serves a similar purpose to the reinforced shoulder mentioned above in that, as the needle enters the material, it creates a larger hole than is needed by the main part of the blade, thus reducing needle-to-fabric friction. The penalty to be paid for this cooler running needle, however, is a tendency to distortion in fine fabrics, and other methods of reducing needle heating are generally preferred.

Scarf

The scarf or clearance cut is a recess across the whole face of the needle just above the eye. Its purpose is to enable a closer setting of the hook or looper to the needle. This ensures that the loop of needle thread will be entered more readily by the point of the hook or looper.

Point

The point of the needle is shaped to provide the best penetration of each type of material according to its nature and the appearance that has to be produced. It is also the part of the needle that must be correctly selected in order to prevent damage to the material of the seam being sewn.

Tip

The tip is the extreme end of the point, which combines with the point in defining the penetration performance.

The features described above are those found on the majority of sewing machine needles.

For use in a particular machine, needles must conform to the machine manufacturer's specification as regards shank diameter, length from butt to eye, and total length. In addition, the different needle manufacturers add their own design features as a result of development work to overcome the problems caused by higher machine speeds and the demand for finer needle sizes to reduce pucker, fabric damage and needle heating effects. Needle strength must remain adequate despite smaller size. Needle thread loop formation is critical if slip stitching is to be avoided, especially on zigzag machines where the hook cannot be set as close as on straight-stitch machines.

2.4.2.2 Problems caused by needle

Needles are available in a wide range of sizes and the choice of size is determined by the fabric and thread combination which is to be sewn. The correct size is essential to good sewing performance but as fabrics tend to become finer and, in many cases, more densely constructed, the demand is for needles and threads that can be used satisfactorily in smaller sizes.

- **If the needle is too small for the thread**, the thread will neither pass freely through the eye nor fit properly into the long groove and will suffer from excessive abrasion as a result. This can lead to costly thread breakages in production because the machine operator must stop to rethread the needle and possibly also to unpick some of the stitching so that a joint does not show in an important part of the garment.

- Even worse, a break in a situation of multi-needle sewing with fabric running through folders could be impossible to repair. The use of too fine a needle when sewing heavy plies of material can lead to the needle being deflected, which can affect the stitch loop pick up and cause slip stitches; it can even lead to needle breakage.

- **If the needle is too large for the thread,** there will be poor control of the loop formation which may cause slipped stitches. There will also be holes in the fabric that are too big for the stitches and give an unattractive seam appearance. An unnecessarily large needle also tends to give rise to damaged fabric along the stitch line and, in closely woven fabrics, pucker along the seam line due to fabric distortion.

- **A needle that is bent** may cause slip stitching if the hook or looper fails to pick up the loop in the needle thread. In a lockstitch this leads to poor appearance and possibly poor strength in the seam. In a chain stitch it can lead to an insecure seam that unravels. The needle can become bent due to faulty operator handling or because the incorrect needle size is used for the fabric being sewn.

- **A needle with a blurred or damaged point** will almost certainly cause damage along the stitch line. On a woven material this will cause fibres or whole yarns to become broken, but because of the nature of a woven fabric there is unlikely to be a complete breakdown of the seam unless the damage is very severe.

- It will show a slightly fluffy appearance along the stitch line. On a knitted fabric, the problem is more serious because a damaged yarn is likely to lead to laddering in the fabric, which is very unsightly and could weaken the seam.

- A yarn in a knitted or a woven fabric can become damaged because it has been broken by the impact of the damaged needle, but in a knitted fabric there is also the possibility that the needle is too large for the size of the knitted loop in the fabric construction and unless yarn can be quickly drawn into the loop from adjacent loops, it will burst.

- Needles can become damaged after striking some part of the machine, usually the throat plate. Occasionally this is a result of the way the operator handles the fabric, but more commonly it arises when the needle becomes deflected during sewing by fabric that is too thick for the size of the needle.

Evidence of the cause of the damage may be present in the form of pock marks on the throat plate where the needle has missed the hole. Beveling out the edge of the hole may help but if the deflection is too severe, a larger size needle may be necessary. Needles can also become damaged as a result of striking a harsh material over a period of time. Material such as the type of denim used for jeans, which is made up in a stiff state and then washed in garment form, may need sewing needles to be changed every 2 h before they become so damaged that they in turn damage the material.

2.4.2.3 Surface treatment for needle

- Nickel plating – This plating is full of corrosion resistance and generally used for the home-use sewing machine.

- Chrome plating – Generally, hard chrome plating is made on the needle, and the needle is superior in heat-proof and wear-proof. The needle is used for the industrial sewing machine.

- Teflon coating – Slide is the best, but durability of coating effect is low.

- Titanium coating – Wear-proof and heat-proof are the best, and this needle is used for extra heavyweight material or the like.

2.4.3 Single needle lockstitch sewing machine working mechanism

- The single needle lockstitch sewing machine is getting drive from the electric motor which is mounted in sewing table. The main parts of the sewing machine are flywheel, eccentric or cam, needle bar, bobbin holder, feed dog and presser foot.

- The flywheel is driven by motor belt and it drives the sewing machine main shaft. Another end of the main shaft is connected with eccentric or cam, which drives the needle bar's up and down movements for every revolution and take-up lever as shown in Figure 2.18.

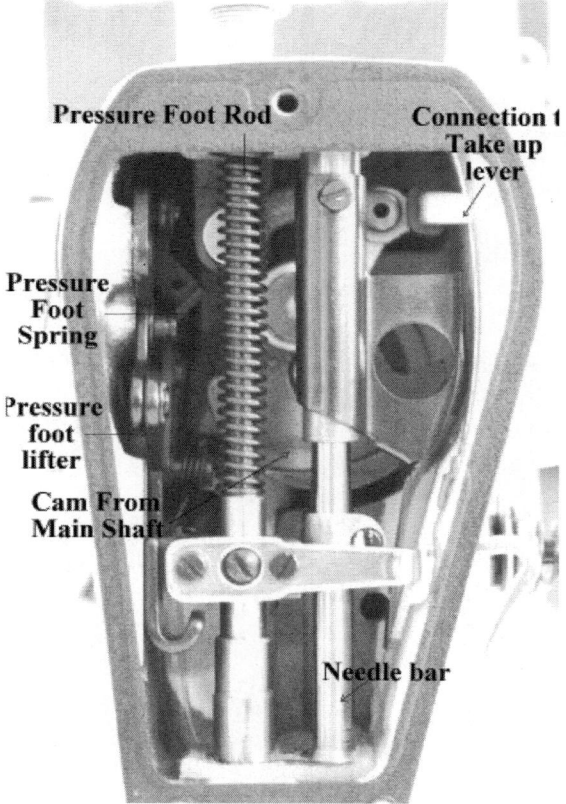

Figure 2.18   Needle bar and Pressure foot mechanism.

- The every rotation of the main shaft moves the needle a full cycle of up and down movements.

- The main shaft is further connected with a bevel gear and two connecting rods to transfer the motion to the bottom of the machine. The bevel gear rod is used to transfer the motion to the bottom centre shaft, which is the main shaft for connecting the bobbin shuttle mechanism. This connection helps the bobbin shuttle to move. The bevel gear mechanism helps to transfer the motion without any energy loss (shown in Figure 2.19).

Figure 2.19   Motion transfer from main shaft to bottom shaft.

- Every one revolution of the main shaft is connected to the every revolution of shuttle hook to make one unit stitch.

- The other two connecting rods from main shaft are used to connect with the feed dog, which has a synchronized movement with needle and bobbin shuttle mechanism.

- The feed dog has to perform four stop motions for every cycle of stitch formation. The feed dog gets drive from main shaft in two ways, one is rotation – to deliver the unit length of the fabric for sewing and another motion is the up and down movements above the throat plate of the sewing machine.

- In Figure 2.20, the rod 1 helps the feed dog to move up and down for the selected distance based on the settings kept. The rod 2 is used to move the feed dog to and fro as required.

- When ever, the stitch regulator is adjusted, the feed dog delivers the required amount of fabric needed by adjusting the feed bracket as shown in Figure 2.21. The revolution per minute of the feed dog is same as the needle stroke.

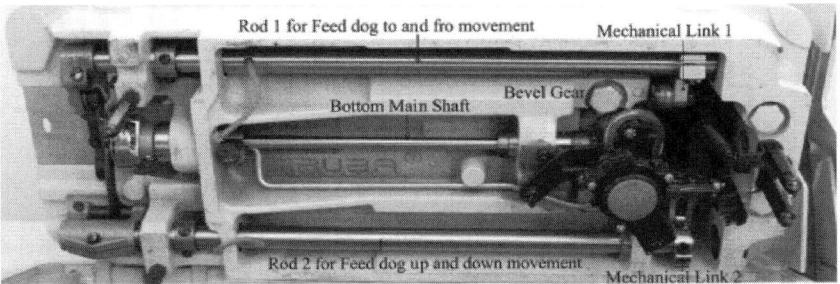

Figure 2.20   Bottom main shaft and mechanism.

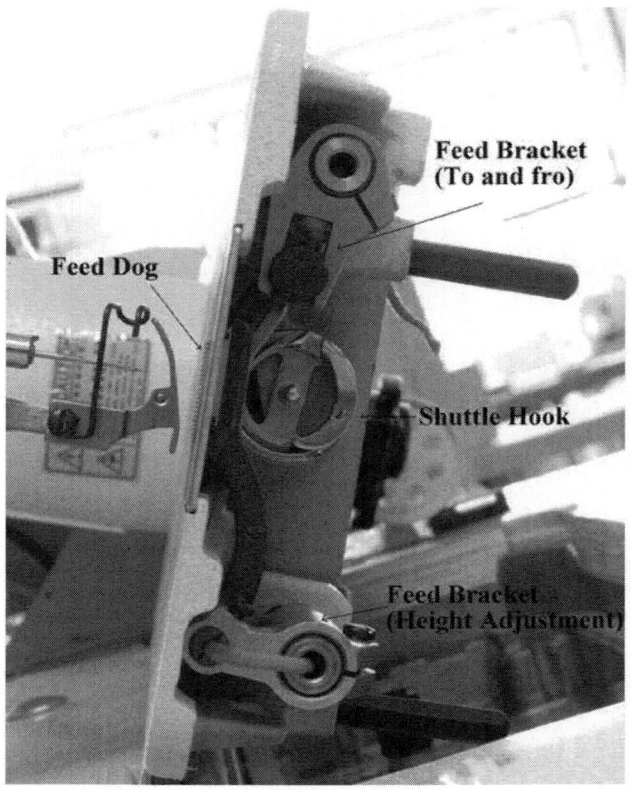

Figure 2.21   Feed dog movement from bottom shaft connection.

- All these motions work together from the single motion of main shaft and help the machine to form the stitch.

2.4.4 Important setting points in single needle lockstitch machine

2.4.4.1 Needle bar adjustment

The functions of the needle bar are,

- Needle bar makes needle up and down, and upper thread penetrates into the material to be sewn.

- Needle bar makes hook or looper scoop the penetrated upper thread.

- Needle bar scoops looper thread at the needle tip (for chain stitch).

The momentum of needle bar is not one kind since the sewing machine sews cloths of various thicknesses. There are three kinds (for heavyweight, medium-weight and lightweight materials) of momentum for 1-needle lockstitch machine. When the needle bar stroke is large, there are such merits as, penetrating force is improved, distance from throat plate to upper dead point of needle tip becomes larger and thick material is easily entered, etc. Demerits are, inertia force is increased and vibration or noise is likely to occur, mechanical load is increased and it is not fit to high speed, needle heat rises, etc.

- The needle bar is marked with few reference lines in the bottom portion. During the initial setting based upon the needle type used, the height of the needle bar is adjusted.

- Needle bar is marked with four reference lines. For general purpose, the bottom most reference line should match the edge of the machine frame. To fix it, the machine pulley needs to be turned to let the needle bar to its lower most position, as shown in Figure 2.22.

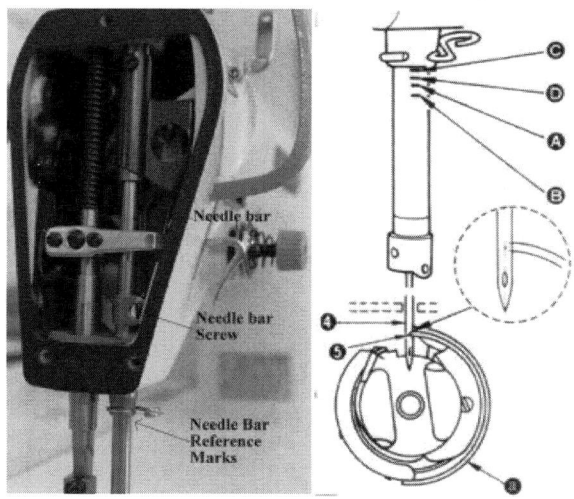

Figure 2.22   Needle Bar setting.

- The needle bar and cam connection screw must be loosened and adjust the height of the needle bar should be adjusted and screw should be tightly secured.

  For DB needle – The marker line A on the needle bar should be aligned with the bottom end of the needle bar lower bushing of the machine followed by tightening the screw.

  For DA needle – The marker line C on the needle bar should be aligned with the bottom end of the needle bar lower bushing of the machine followed by tightening the screw.

2.4.4.2 Presser foot height

- The standard height of the pressure foot is 2 mm–6 mm when the presser foot is raised by means of presser foot lifter.

- To adjust the presser foot height, the pressure applied on the presser foot is released by loosening the presser foot pressure regulation screw (3 and 2) as shown in Figure 2.23. This will reduce the pressure applied on the presser foot.

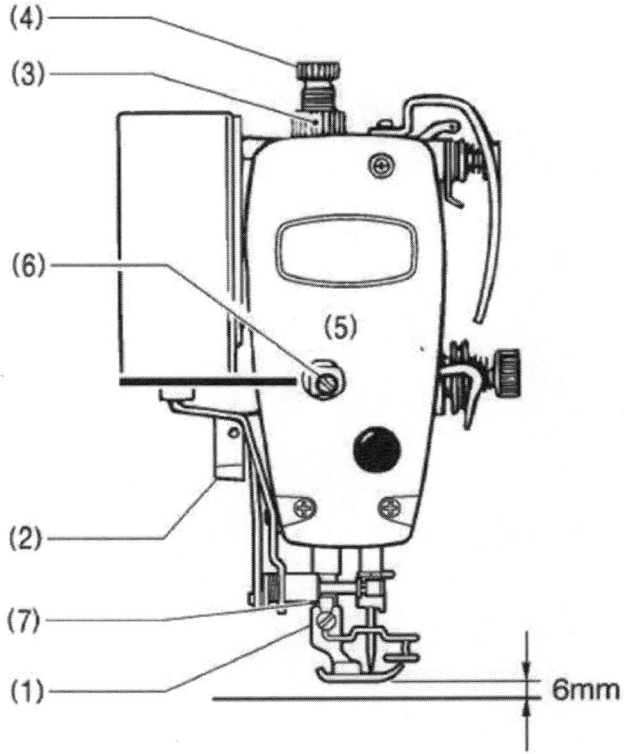

Figure 2.23   Pressure foot Height adjustment.

- After that, by raising the presser foot lifter (2), the presser foot is raised (1). Then by removing screws (5 and 6), the height of the presser foot bar (7) can be adjusted and screw should be securely tightened.

2.4.4.3 Feed dog height and angle adjustment

Height adjustment

The main functions are: it makes the sewing product move per stitch, used to change amount to move and forms stitches suitable for the sewing product.

The feed dog height from the needle plate depends upon the type of fabric used.

1. Normal fabric – 0.75 mm to 0.85 mm

2. Heavy fabric – 1.15 mm to 1.25 mm

3. Light weight fabric – 0.7 mm to 0.8 mm

   • The feed dog height is adjusted by loosening the screw in the feed dog connection rod and height is adjusted by moving the feed bracket up and down followed by tightening the screw securely (Figure 2.24).

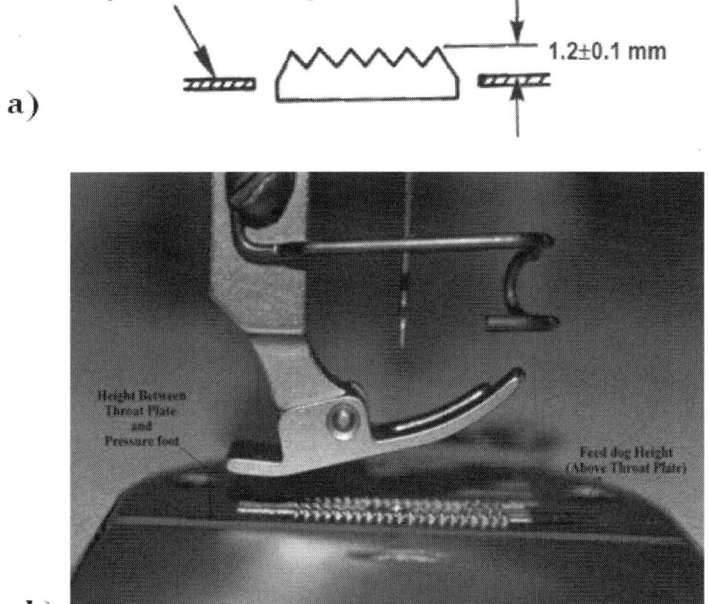

Figure 2.24 (a)Feed dog height and (b) Pressure foot lift.

Angle adjustment

The fabric puckering can be prevented by loosening the height adjustment rod screw and then turning the feed dog shaft in the direction of arrow mark mentioned. This process makes the feed dog to tilt in the front side to up.

   • To prevent the fabric from uneven feeding, the height adjustment rod screw should be loosened and the feed dog shaft is turned in the opposite direction of arrow mark mentioned. This process makes the feed dog to tilt in the front side to down (Figure 2.25).

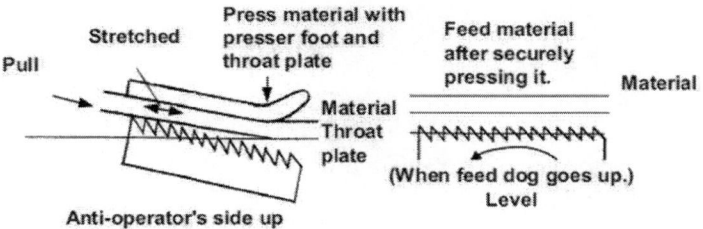

Figure 2.25 Feed dog tilt.

Feed dog pitch

The pitch length of the feed dog is the distance between the successive teeth (Figure 2.26a).

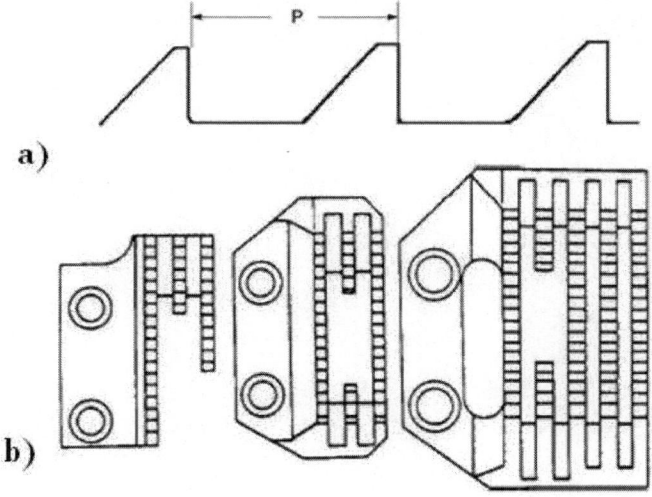

Figure 2.26 (a) Feed dog pitch length and (b) rows.

Slim pitch: This is suitable for lightweight and soft materials. If this pitch is used for heavyweight materials, bite to materials is deteriorated and feed force becomes insufficient.

Coarse pitch: This is suitable for thick and hard materials in some degree. If this pitch is used for lightweight material sewing, it will be a cause of puckering. For lockstitch: Slim pitch 1.15 mm, standard 1.5 mm, coarse pitch 1.8 mm and coarse pitch 2.0 mm; for extra heavyweight materials: 2.5 mm–4.5 mm

Number of teeth (rows)

The less the number of teeth (rows) is, the better the sharp curve stitching becomes (Figure 2.26b). The more the number of teeth (rows) is, the better straight feeding, feed force and stability of materials become. When feed dog is located at this side of hole of throat plate, feeding to overlapped section and bite at the start of sewing are improved.

For the elastic materials such as knit or the like, it is likely to be good to feed materials at the front or rear of needle entry. If there is no feed dog on this side, the material is in the state that it is pulled by the feed dog located in the rear of needle, and the material is sewn while it is somewhat stretched.

2.4.4.4 Needle to shuttle hook timing setting

- To set up the shuttle hook and needle pick up timing, the screw which connects the bottom main shaft and shuttle hook should be loosened.

- After correcting the needle bar height, the flywheel is rotated manually to allow the needle to go to their bottom most position (Figure 2.27).

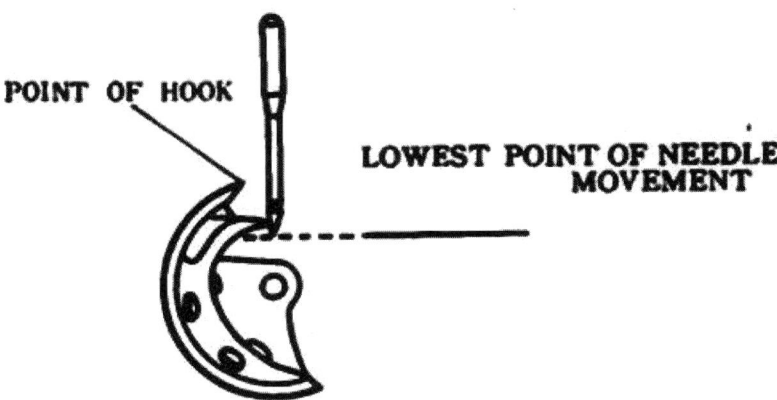

Figure 2.27   Needle to Hook setting.

- The shuttle hook is kept at the needle movement path and flywheel is slowly rotated to make the needle to rise from bottom to top position.

- The needle and shuttle hook is aligned to meet at the centre of needle scarf position.

Here in Figure 2.28(a) the needle aligned with the shuttle as such the hook of the shuttle exactly meets the scarf of the needle, (b) Blade point of outer hook catches upper thread, (c) Upper thread separated by inner and outer hook movement (d) Immediately after the thread passes through the inner hook the take up lever stats working on tightening the thread, so that thread comes off the hook under the fabric and tighten the stitch against the looper thread.

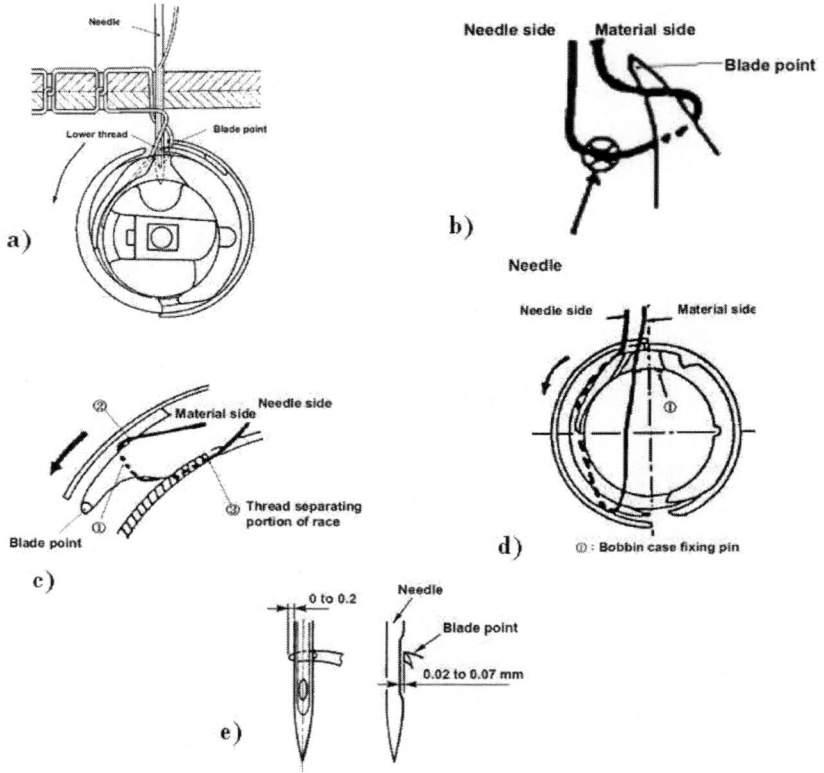

Figure 2.28   Needle to Hook setting.

- It is the closest setting in the machine. The distance between needle and shuttle hook is barely less than 0.5 mm (Actual clearance is 0.02 mm–0.07 mm).

- Once the setting is done, screw is securely tightened in the main shaft, without altering the needle and shuttle hook position.

2.4.4.5 Rotary hook

Figure 2.29 represents the diagram of a rotary hook. The major parts of the hook are,

Locating tab notch – It helps the locating tab to loosely fit on the top edge and thus prevent the movement of the bobbin case during sewing process

Thread loop pusher – It helps the thread to form loop during its circular motion

Bobbin shaft – It is used to fix the bobbin case

Hook basket – It provides clearance at the bottom of the rotary shaft and bobbin case, it helps the thread to pass out and form the stitch at required time

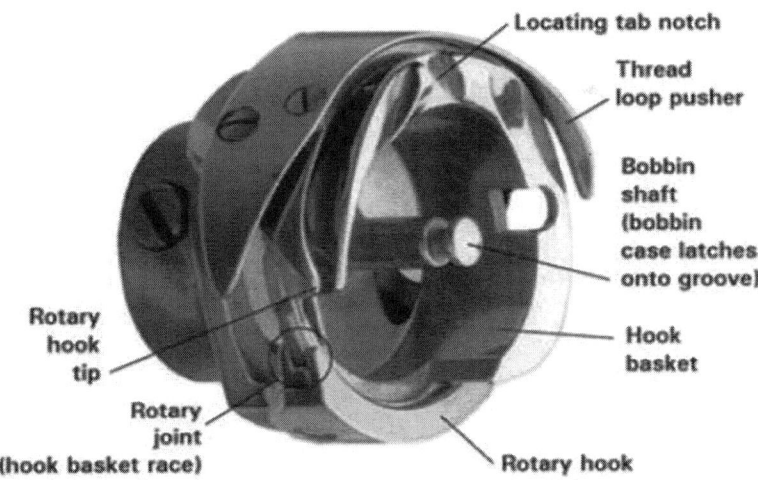

Figure 2.29   Hook Set parts.

Rotary joint – It is called as hook race and it helps the hook basket to allow the thread to run

Rotary hook tip – It helps in picking the thread during every revolution

Working of rotary hook

- The rotary hook is attached with the bottom shaft of the machine directly and synchronized with the movement of the needle.

- The bobbin is held in the shaft in a bobbin case, which is fixed with the rotary hook by rotary joint. This helps the stitch formation process by rotating the bobbin along with the rotary hook without moving the bobbin case.

- Since the bobbin case is not rotating during sewing, it falls down and interlocks with the rotary joint and so the bobbin case cannot move from its position. Once the rotary tip picks the thread, the rotary hook passes the rear of the loop of top thread behind the bobbin and bobbin case (Figure 2.30).

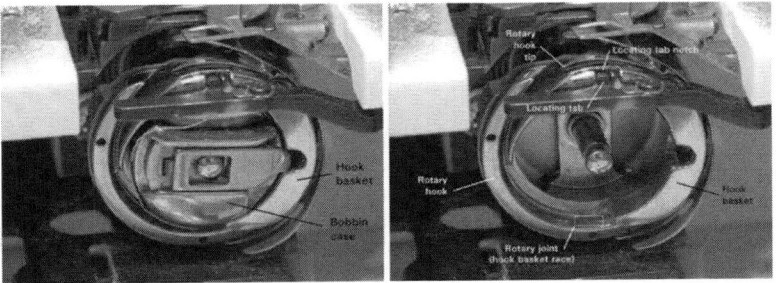

Figure 2.30   Role of Hook set in stitch formation.

While the full revolution is completed, the rotary joint provides the space between bobbin and bobbin case. It provides the clearance for the bobbin thread to form the stitch.

2.4.4.6 Timing diagram of sewing process

The timing diagram of sewing process is given below. It gives the details about movement of sewing elements at various degrees to form a stitch, as shown in Figure 2.31.

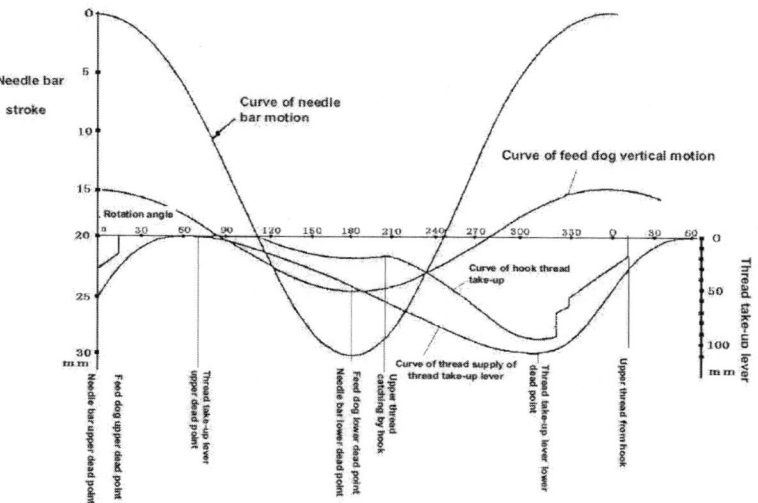

Figure 2.31   Timing Diagram of Single needle lock stitch machine.

- 0°/360° = The needle will be at throat plate position.

- 90° = Needle bar will be at top most position. The bobbin will start its activity and the bobbin hook will come near to throat plate position.

- 180° = The needle will come down with needle thread. The bobbin thread will come up and form the stitch at the stitch form line.

- 270° = Needle bar will be at lower most position. Loop will be formed fully and the bobbin will move forward to complete its rotational cycle.

- 360° = The needle will move up from the bottom to throat plate position and the bobbin will finish its full circle and reach its rest position.

2.5 Stitch-formation in single needle lockstitch machine

The stitch-forming elements are the mechanical parts which on correct synchronization, form stitches and sew seams or stitching. Many are specific to the stitch or machine type. The various stitch-forming elements are

- Thread control devices, which include tensioners and take-ups
- The needle
- The feed dogs
- Throat plate, tongues and chaining devices
- The presser foot
- The rotary sewing hook, loopers and spreaders
- Bobbin and bobbin case

Correct stitch formation is dependent on the appropriate combination of these parts and their precise synchronization. Good maintenance is essential for maintaining the required precision.

2.5.1 Thread control devices

- Thread control devices include **thread guides, tension devices and take-ups**, which are necessary to provide uniform thread flow.

- Thread guides control the positioning and movement of thread. Damaged or faulty thread guides can damage sewing thread and cause thread breakage and weak seams.

- Tensioning devices control the flow and tightness of thread as it travels through the stitch-forming parts of a sewing machine. Tension determines the balance and tightness of a stitch. Tensioning devices consist of a pair of tension discs, a spring and thumb nut that can be adjusted to control the ease with which thread passes between the discs. In stitch formation, excessive tension may cause thread breakage or puckered seams; too little tension allows slack and excessive looping, and loose stitches that will not hold. Tension needs to be adjusted when thread, fabric, and/or needles are changed.

- Thread take-up controls the supply of thread required to form each stitch. It gives extra thread to the needle to form the stitch but takes it away to set the stitch.

2.5.2 Needles

Needles carry the thread through the fabric so a stitch can be formed. Proper stitch formation is dependent on the formation of a loop of thread below the fabric that can be picked up by a hook or looper.

2.5.3 Lower stitch-forming devices

Lower stitch-forming devices include the hook found on bobbin cases of lockstitch machines and loopers. A hook is a rotating device encompassing the bobbin case that picks up the needle thread loop to form a lockstitch. If the hook misses the loop, no stitch is formed. Precision and timing are critical to stitch formation.

Loopers may carry lower threads to interlock with needle threads or other looper threads. Loopers have a set motion pattern that must be synchronized with the needle motion and feeding. Spreaders work in conjunction with a looper to assist loop formation. They move the thread but do not carry the thread. Threading fingers move the cover threads back and forth in forming the cover stitches.

2.5.4 Stitch tongues or chaining plates

They are pointed metal extensions that may be part of or attached to throat plates. Stitch tongues are essential for the formation of the three-dimensional stitches. Tongues vary in length and shape depending on the requirements of the specific stitch type and operation. A long, thin pointed tongue is needed for tight, close stitching such as a rolled edge, while a flatter, wider tongue is used for flat seaming.

2.5.5 Stitch formation

Step 1: The first step in producing a stitch on a sewing machine is the formation of the needle thread loop. This step is the same regardless of the type of stitching being produced, or the nature of the machine being used. Failure to generate a good loop will cause many different problems such as skipping, breaking thread, loose stitches, threading bunching up, etc.

Proper formation of this loop depends on the tendency of the thread to bulge away from the needle as it is drawn upward after reaching the lowest point of its stroke – due to inertia and friction against the material through which it passes.

Step 2: The fabric has two threads running through it. One thread runs through the needle from a spool and the other comes from the bobbin. The needle lowers and penetrates the fabric, carrying its thread with it. The loop formed in the needle-thread is then entered by the point of the sewing hook (Figure 2.32a).

Step 3: The needle begins to lift and forms a loop of thread on the fabric's under side. As the hook case turns, the hook approaches, catching the looped thread. As the needle continues to rise and the hook progresses in its rotation, the needle-thread take-up arm provides sufficient slack thread to be drawn down through the fabric to increase the size of the loop (Figure 2.32b).

Step 4: The needle rises above the fabric. As the hook case continues to turn, it carries the loop of thread. On its first rotation, the sewing hook carries the needle-thread loops around the bobbin case and bobbin, the inside of the loop sliding over the face of the bobbin-case while the outside passes around the back, to enclose the bobbin-thread (Figure 2.32c).

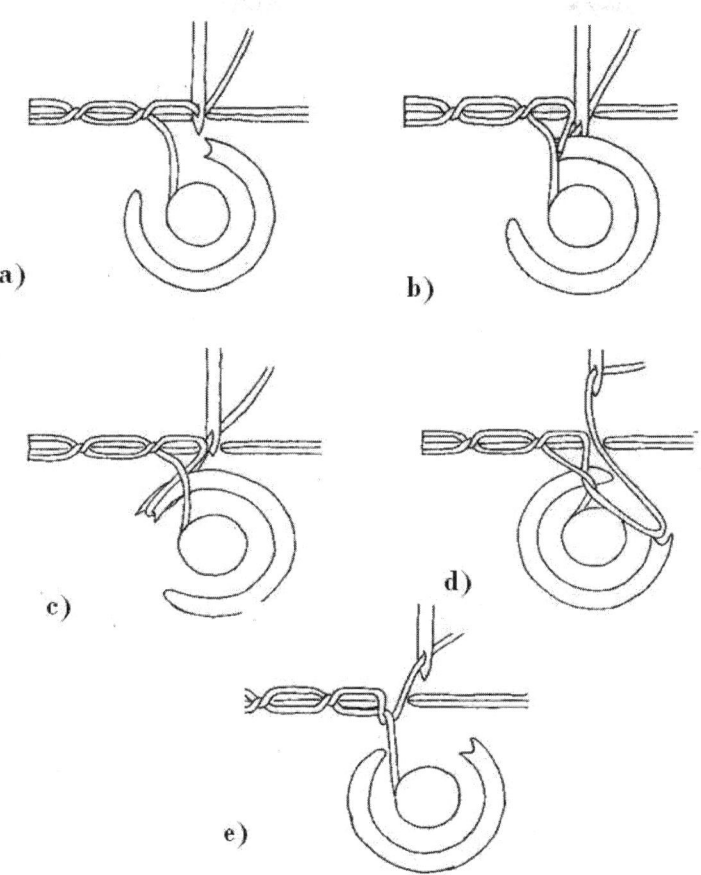

Figure 2.32   Needle and Hook set movemets.

Step 5: The needle is still above the fabric. The hook case finishes carrying the loop over the bobbin. As the needle-thread take-up starts to rise, the loop is drawn up through the 'cast-off' opening of the sewing-hook before the revolution is complete (Figure 2.32d).

Step 6: The needle's thread is pulled taught by a lever upstream (not visible in the diagram). This action pulls the loop from the hook and thus completes the stitch. The stitch is now ready to be repeated (Figure 2.32e). During the second revolution of the sewing-hook the thread take-up completes its upward stroke, drawing the slack thread through the material and settling the stitch. Meanwhile, the feed dog has moved forward carrying the material with it and drawing the required length of under-thread from the bobbin. The presser-foot guards against the slippage by holding the fabric firmly against the teeth of the feed dog while the feed dog is carrying the fabric across the smooth face of the throat plate or needle plate. There are many factors that can prevent the correct formation of the stitch, namely, they can be:

(i)   Anything which inhibits the formation of the needle-thread loop

(ii)   Anything which prevents a correctly formed loop from being entered by the point of the sewing hook

(iii) Anything, which interferes with the free running of either, the needle-thread or the bobbin thread

(iv)   The size and type of the needle and sewing thread

(v)   The type of fabric and the manner in which it is fed across the throat plate of the machine

2.6   Working mechanism of double needle

• The DNLS sewing machine (shown in Figure 2.33) is getting drive from the electrical motor which is mounted on the sewing machine table. The main parts of the sewing machines are fly wheel, eccentric or cam, needle bar, bobbin holder/shuttle holder, feed dog and presser foot. The working mechanism of the double needle lockstitch machine is very similar to the single needle.

Figure 2.33   Double needle lock stitchmachine.

• The fly wheel is driven by motor belt and the main shaft is connected to the flywheel. The main shaft is connected to the cam that drives the needle bar and take-up lever (Figure 2.34). Every rotation of the main shaft results in one up and down movement of the needle. In the same way the stitch forming mechanism also has two bobbins hook set mechanism on both sides of the feed dog.

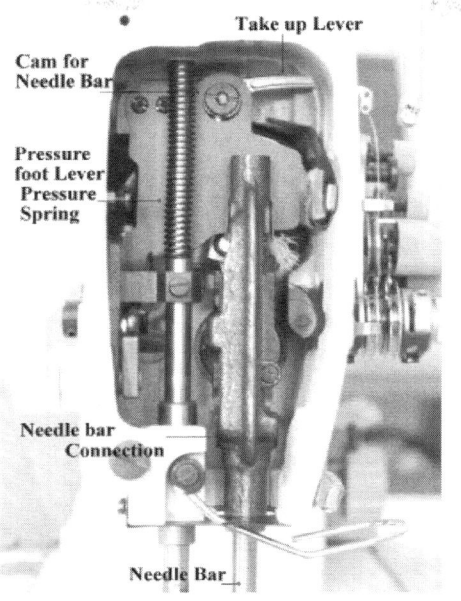

Figure 2.34 Double needle Machine needle bar mechanism.

- Hence the needle needs to be fixed in such a way that the scarf portion of the needle should face away from the machine (Figure 2.35).

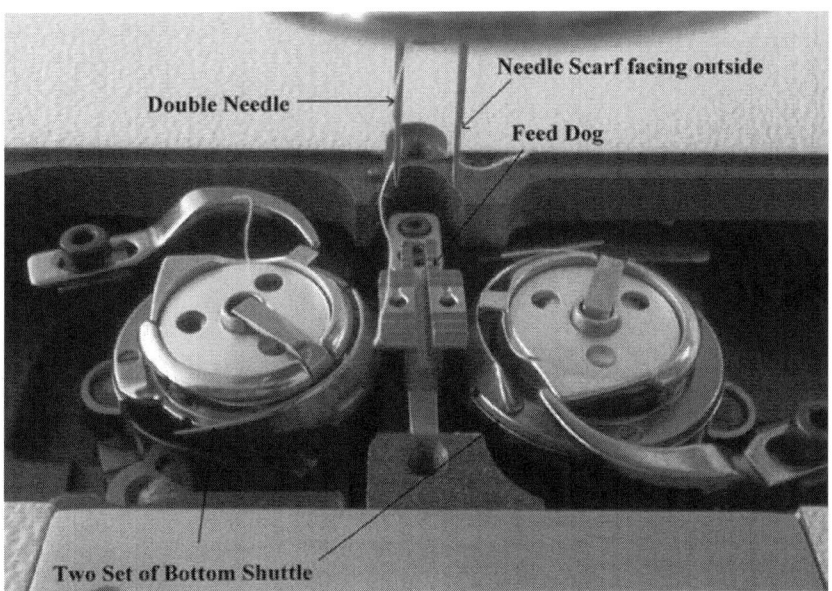

Figure 2.35 Double needle lock stitch machine hook set (Horizandle set up).

- The motion from the main shaft is transferred to the bottom shaft using a belt. The belt transfers the motion to the bottom shaft, which has the connection with both the bobbin hook/shuttle set as shown in Figure 2.36.

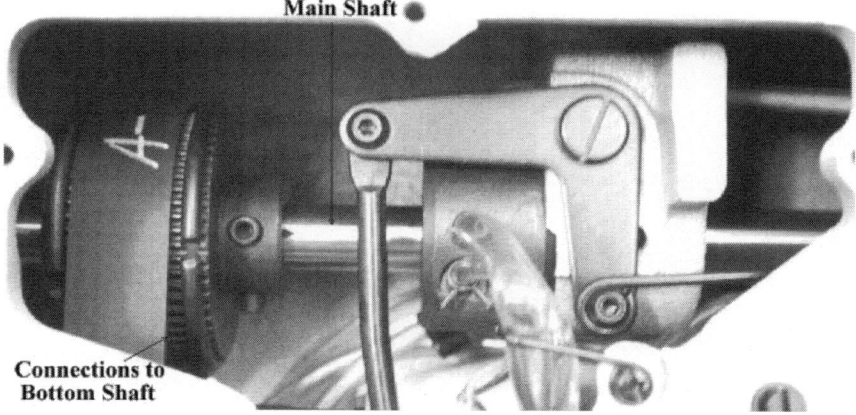

Figure 2.36   Main shaft to transfer motion from flywheel to bottom shaft.

- The feed dog gets drive from main shaft to move in two ways: one motion being to and fro to deliver the unit length of fabric to required length and the other being the helical movement of feed dog to move over the throat plate. The main rod transfers the full motion and keeps the bobbin hook set synchronized with needle movement (Figures 2.37 and 2.38).
- The adjustment in the stitch regulator adjusts the fly time of the feed directly. Hence, the adjustment of feed dog always leads to alteration of fabric length delivered. The selection of feed dog will depend on the stitches per inch and requirements.

Figure 2.37   Bottom shaft and Supplementary shaft for double needle lock stitch machine.

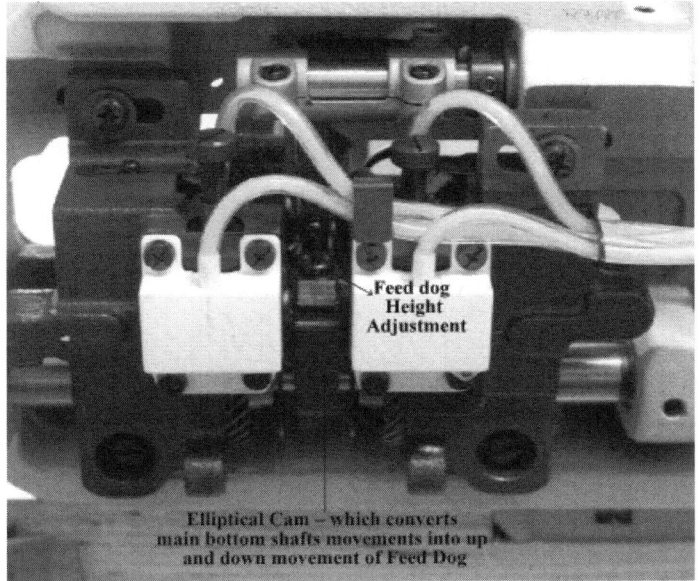

Figure 2.38   Double needle machine Driving mechanism of hook set assembly.

2.6.1 Important setting points in double needle lockstitch machine

2.6.1.1 Needle to hook setup

- The needle is fixed in the needle bar by facing the scarf of each needle opposite (outside) to each other. The needle is kept at the bottom most position

- Manually the needle is moved from bottom most position to the top position by rotating the flywheel. When needle scarf passes the shuttle hook, each shuttle hook is manually moved towards the needle and fixed with the hook passing through the scarf of the needle on both the sides.

- The shuttle hook screw is tightened securely without altering the position as shown in Figure 2.39. The clearance between the needle plate and rotary hook is approximately 0.9 mm–1.2 mm.

- Then the machine is slowly run manually after threading to check whether the shuttle picks the thread from the needle during every cycle.

- If the pickup is not done, may be the distance between the hook and needle scarf must be high or the pickup may be too early or late. If the thread is not picked properly, it might lead to poor stitching performance.

Screw for feed dog height adjustment

Figure 2.39   Double needle machine hook set assembly adjustment Screw.

2.6.1.2 Presser foot and feed dog height

- The distance between the needle plate and presser foot is approximately 7 mm and this can be adjusted by moving the presser foot bar, as mentioned earlier in single needle machine section.

- The feed dog height can be adjusted by altering the screw shown in Figure 2.39.

2.7   Overlock machine

An overlock stitch sews over the edge of one or two pieces of cloth for edging, hemming or seaming. Usually an overlock sewing machine will cut the edges of the cloth as they are fed through (such machines are called 'sergers'), though some are made without cutters. The inclusion of automated cutters allows overlock machines to create finished seams easily and quickly. An overlock sewing machine differs from a lockstitch sewing machine as it utilizes loopers fed by multiple thread cones rather than a bobbin. Loopers serve to create thread loops that pass from the needle thread to the edges of the fabric so that the edges of the fabric are contained within the seam. Overlock sewing machines usually run at high speeds, from 1000 rpm to 9000 rpm, and are used mostly in industrial setting for edging, hemming and seaming a variety of fabrics and products.

Overlock stitches are extremely versatile, as they can be used for decoration, reinforcement or construction. Overlocking is also referred to as 'overedging', 'merrowing' or 'serging'. Though 'serging' technically refers to overlocking with cutters, in practice the four terms are used interchangeably.

2.7.1 Selecting an overlock machine

The type of overlock machine used will determine the kind of sewing technique that can be used. A two-thread model is the best for finishing seam edges. It can also be used to stitch flatlock seams. A flatlock seam is usually sewn with the wrong sides of the fabric together. The seam is then pulled so the seam allowances slip and lie flat inside the stitching. This seam is useful when seaming sweatshirts, jogging suits and other active wear made from sportswear fleece and velour. It can be used to stitch elastic and lace to lingerie. Hems can also be sewn using some two-thread overlock machines.

A three-thread overlock trims, stitches and overcasts seams in one operation. The three thread overlock seam is most useful when sewing knits. It can be used to sew woven fabrics, but not in areas that will receive a lot of stress. The three-thread can also be used to sew pin tucks, make narrow rolled hems and to finish fabric edges. Decorative stitches are possible using a variety of threads or yarns, such as crochet yarn, pearl cotton, buttonhole twist, lightweight knitting yarn, narrow ribbon or metallic thread. Some three-thread models can be converted to do two-thread stitching.

The four-thread overlock will stitch a chain stitch or a safety stitch as it stitches and overcasts seams. The chain stitch model is most useful when seaming woven fabrics. The safety stitch model can be used to sew woven or knitted fabrics. By using one needle and one looper, the four-thread model is easily converted to do two-thread stitching. Some four-thread overlock machines are converted to do three-thread stitching.

Two-thread overlock stitch

- Used for finishing seam edges when seams are to be sewn using a conventional machine.

- Used to create the flatlock seam in loosely fitting active sportswear, knits and some trendy fashions in woven fabrics.

- It should be avoided in areas that will receive stress.

Three-thread overlock stitch

- Suitable for sewing woven and knitted fabrics, but should be avoided in areas of stress.

- Can be used when sewing most shirts, blouses, skirts, dresses, lingerie and swimwear.

- It should be avoided in areas of stress which include crotch seam of pants and sleeve seams of shirts or blouses.

Four-thread overlock with safety stitch

- Used for same projects as the three-thread overlock stitch, but can be used in areas that receive stress.

- Suitable for sewing blouses, shirts, skirts, dresses, pants, lingerie, action wear, swimwear and sleepwear in knitted or woven fabrics.

Four-thread overlock and chain stitch

- Suitable when sewing woven fabrics, even in areas of stress.

- Stitch does not stretch so should not be used in knitted garment when stretch is required in seams.

- Used when sewing shirts, blouses, skirts, pants, sleepwear and draperies.

- Use chain stitch when fitting garments, since stitches are easily removed. The stitches are shown in Figure 2.41.

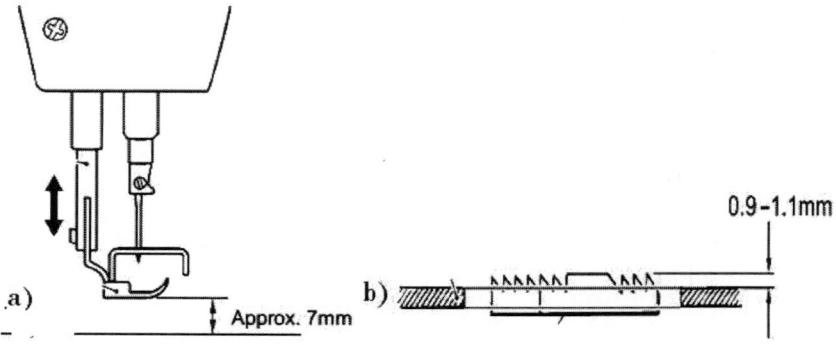

Figure 2.40   Needle bar and feed dog height.

2.7.2 Special features of overlock machines

There are a number of features to consider carefully when selecting an overlock machine as shown in Figure 2.41. The major concern for users is the threading process. It is to be noted that the machine has a coded threading system – colours, symbols or numbers may be used. The thread stand should have adaptors for the use of cones or spools of thread. However, the use and care manual can be used for obtaining threading instructions. The loopers are the most difficult parts to thread. A special threading tool and a pair of tweezers are usually provided. The lower looper is usually threaded before the upper looper.

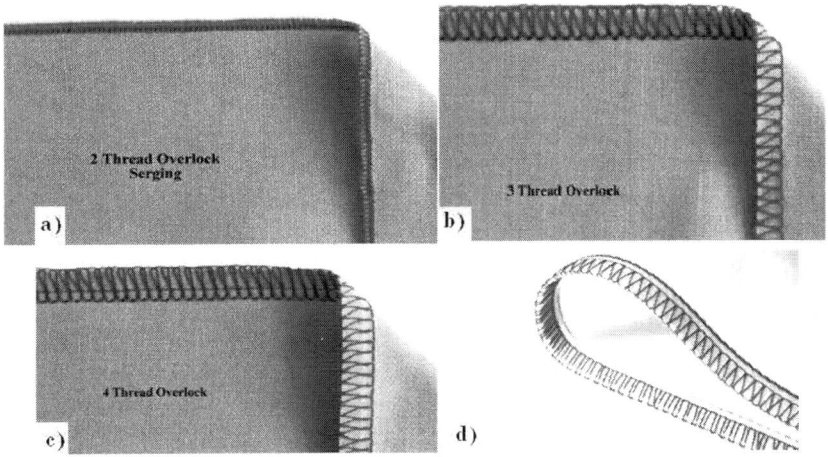

Figure 2.41   Overlock stitch (a) two thread, (b) three thread, (c) four thread safety stitch and (d) four thread chain stitch.

2.7.3 Over lock machine – special parts

Overlock sewing machines will trim, stitch and overcast seams as they sew, and they sew faster than conventional sewing machines, up to 1500 stitches per minute. Because of these features, an overlock can save time and give a professional appearance to items constructed. Decorative effects can be achieved by using special threads in the loopers. Threads, such as metallics, silks and pearl cotton, can be used to sew seams or to finish garment edges. For special effects, fine knitting yarn, buttonhole twist, crochet thread and even narrow ribbon can be used.

Overlock sewing machines are specialized, but they do not replace the conventional sewing machine. Overlock forms interlocking stitches using one or two needles and one or two loopers. The overlock stitch is more like crochet or knitting than the stitch of a conventional machine. Loopers replace the traditional bobbin and interlock threads together. Seam edges are trimmed by blades located just in front of the needle(s). In some models, the upper knife can be rotated up to disengage the cutting action. In other models, one blade must be removed when trimming is not desired.

2.7.3.1 Loopers

Loopers are dull pointed metal piece which have definite motion cycle and houses the looper point to grasp the needle thread as looper moves towards its pointed end. The loopers pointed end grasps and pulls the needle thread from the needle eye as the needle rises.

Loopers are two types

1. Eye looper (Figure 2.43a) – Which carries a thread whose function is to interlock with needle or looper thread (400, 600)

2. Blind looper (Figure 2.43b) – Which does not carry a thread of their own (100, 101, 102, 500)

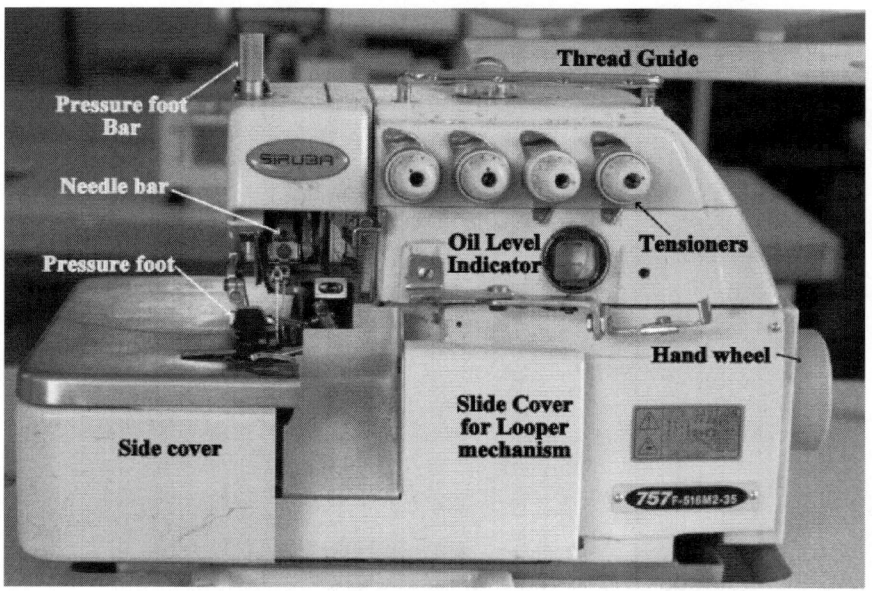

Figure 2.42   Overlock machine Parts.

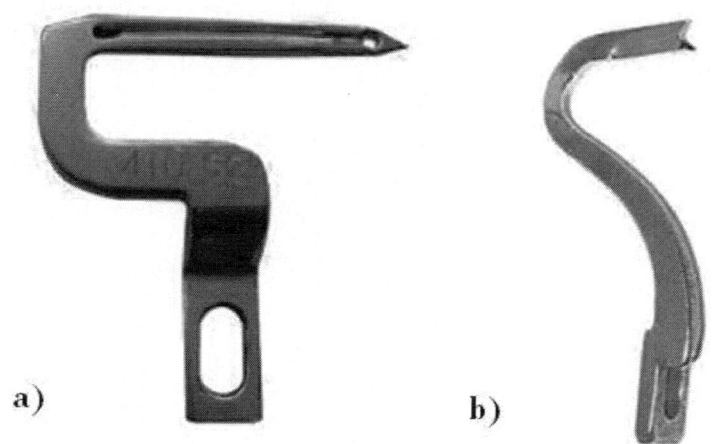

Figure 2.43   Overlock sewing machine loopers (a) Looper with eye (b) Blind looper.

2.7.3.2 The tension dials

The tension dials may be marked with numbers to indicate degree of tension (shown in Figure 2.44), but some machines use + or − signs. The tension system differs from machine to machine. On some machines, tension dials are sensitive, requiring only a slight turn to adjust the flow of thread. On others, dials must be rotated several times before the thread tension is affected.

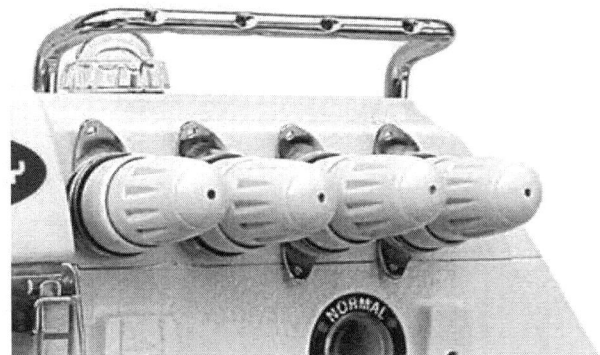

Figure 2.44   Overlock sewing machine tension dials.

2.7.3.3 The presser foot

- The presser foot of an overlock machine is much larger than that of a conventional machine (Figure 2.45)

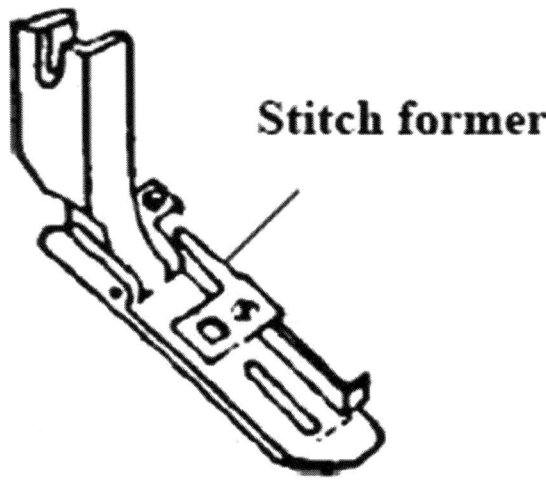

Figure 2.45   Overlock sewing machine presser foot.

- The pressure on most models can be adjusted to the weight of fabric being sewn. Because the foot is so long, there is seldom a problem of fabrics feeding unevenly as they are sewn

- On some models

2.7.3.4 The throat plate

- The throat plate or presser foot will have a special stitch former not found on a conventional machine as shown in Figure 2.46.

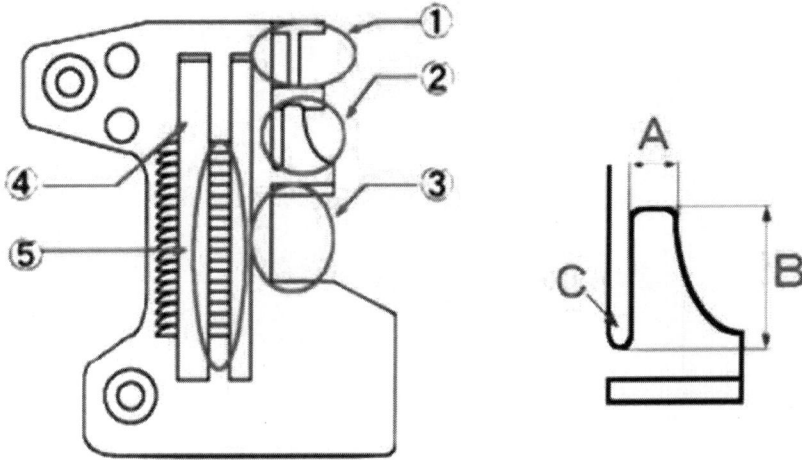

Figure 2.46   Overlock sewing machine Throat plate.

- On some machines, this former can be adjusted to change the stitch width.

- On others, the throat plate or presser foot must be changed when a different stitch width is desired. Stitch widths range from less than 1 mm to greater than 5 mm.

(1) **Location of the auxiliary feed dog teeth**

   • The auxiliary feed dog teeth helps produce chain-off thread with consistency.

(2) **Overedging width**

   • The over edging width differs with the width of the over edging claw.

   A – The claw width determines the overedging width.

   B – Thread tension varies according to the entire length.

C – Needle entry.

(3) **Location of a cloth cutting knife**

• The fabric is cut here. The cutting width is determined by the location of the knife.

(4) **Feed dog groove**

• The feed dog for the single needle overlock machine comes in three different types, i.e., feed dog with 1 row, 2 rows and 3 rows of teeth.

• Efficiency of feed and straight feeding capability of the feed dog increases with the number of teeth.

• The feed dog with single row of teeth is characterized by its ability to support sharp-curve sewing.

(5) **Throat plate slot**

• The slot prevents the feed mechanism from returning (in prevention of an uneven pitch). The slot also works to improve the differential effect.

2.7.3.5 Cutting knives

• The cutting knives are made to last for many hours of sewing (Figure 2.47), it is important to determine how easy they are to change. One blade will need to be replaced more often than the other; it is the less expensive of the two blades.

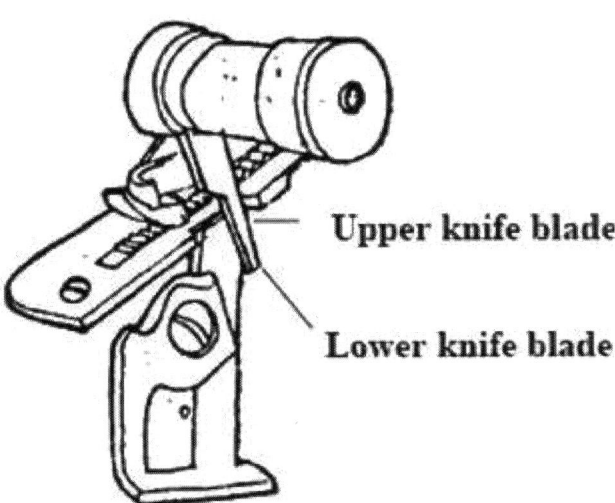

Figure 2.47 Overlock sewing machine Cutting Knife.

- On some machines, the knife position can be adjusted to vary the seam allowance width. Some knives can be disengaged so that trimming does not occur.

2.7.3.6 Pneumatic flat cutter

This cutter sucks chain-off thread through a chain-off thread suction opening which is located parallel to the throat plate to allow trimming. This cutter is capable of trimming chain-off thread produced on the double chain stitch side of a safety stitch machine. Because of this feature, the pneumatic flat cutter is mainly mounted on the safety stitch machine as shown in Figure 2.48.

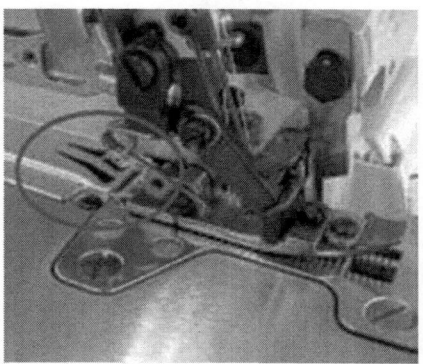

Figure 2.48   Overlock sewing machine pneumatic flat cutter.

2.7.3.7 Pneumatic side cutter

This cutter trims chain-off thread on the side behind the presser foot as in Figure 2.49. The cutter leaves shorter chain-off thread on the material and therefore is suited for blind-hemming and serging of an endless work piece. It is mainly mounted on the overlock machine.

Figure 2.49   Overlock sewing machine pneumatic side cutter.

2.7.3.8 The chain-off thread suction

This unit sucks chain-off thread by means of compressed air. Another type which is capable of simultaneously sucking thread chips produced after trimming to help improve the work environment is also available as shown in Figure 2.50.

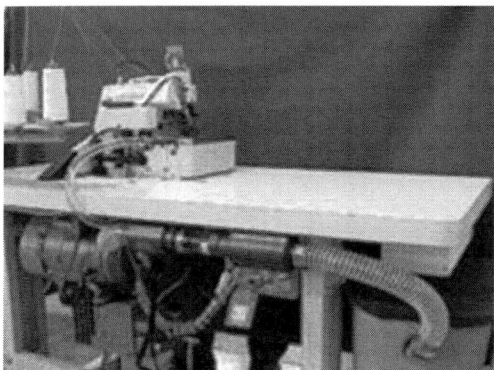

Figure 2.50   Overlock sewing machine pneumatic chain off thread suction.

2.7.3.9 Chain-off thread trimming device

This type of chain-off thread trimming device is the most frequently used one. The thread trimmer is located at the rear of the throat plate and the knife comes down from above. Because of this mechanism, this thread trimming device is called guillotine type.

2.7.4 Working mechanism of overlock machine

- The overlock sewing machine gets drive from the electric motor which is mounted in sewing table. The main parts of the sewing machine are flywheel, main shaft, needle bar, loopers, feed dog and presser foot.

- The flywheel is driven by motor connected by belt and it drives the sewing machine main shaft. Another end of the main shaft is connected with eccentric or cam, which drives the feed dog (stationery) movement.

- The needle bar's up and down movements are controlled by a link mechanism driven from top side of the main shaft and which is synchronized with the feed dog movement for every revolution.

- Other important mechanism used in the chain stitch formation is loopers. There are three important loopers, namely, top looper, bottom and chain stitch loopers. They get drive from the main shaft directly through a link.

2.7.4.1 Important setting points in overlock machine

2.7.4.1.1 Needle bar motions

- There are two types of needle bar set up available in overlock machine generally. On straight needle machine, needle bar is generally less inclined backward at around 20°–23° from vertical or the needle bar is just fixed exactly perpendicular to the feeding direction as shown in Figure 2.51. In all the cases, the motion of the needle is in a straight line.

- In order to form the overlock seam, the lower looper passes behind the needle and moves to the right. The upper looper passes behind and the lower looper moves upward, passing in front of the needle.

- If the needle moves up and down vertically, the upper looper will not be able to move in front of the needle, so the needle bar in the overlock machine leans back at an angle to provide enough space for the width of the upper looper movement.

Figure 2.51   Overlock sewing machine Needle bar.

- The needle bar may be one needle or two needle type, in the case of 2-needle machine the spaces between the needle is around 2 mm, 2.5 mm or 3 mm.

2.7.4.1.2 Type of looper movement

- The lower looper moves under the fabric from the left side of the needle to the right side of the needle while passing behind the needle. At the edge of the fabric, it crosses the upper looper and passes its front.

- The upper looper moves from the right end of the fabric and it crosses the lower looper.

2.7.4.2 Different types of feeding mechanism

- The fabric is gripped between the feed dog and presser foot. It is fed a pre-determined distance; this movement is a combination of vertical and horizontal motion. Feed dogs may be divided into two types. Single feed dog and double feed dog system. In the second type, one feed dog is in front of the needle and the other one behind the needle. The choice is based on the application.

Differential feed dog is used for creating gathering or stretching of fabric as they are sewn up under the needle. For every 1 cm material feed, 2 cm of the material is delivered in the case of stretch effect. This is generally used for the woven fabric. In the other case every 2 cm of the fabric fed, 1 cm of the material is delivered which gives an effect called gathering which is generally used in the knitted fabric.

To adjust the differential feed device, following instructions need to be adhered to

Figure 2.53(a) shows the adjustment screw for stitch length regulation. By turning the screw, the movement distance of the feed dog can be altered.

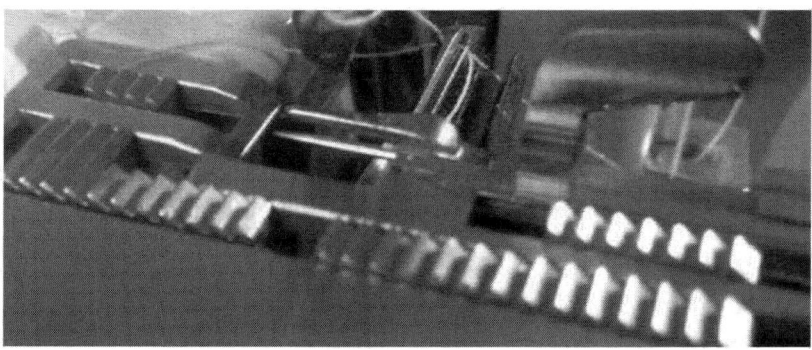

Figure 2.52   Overlock sewing machine – 2 set of feed dog for differential feed.

a)

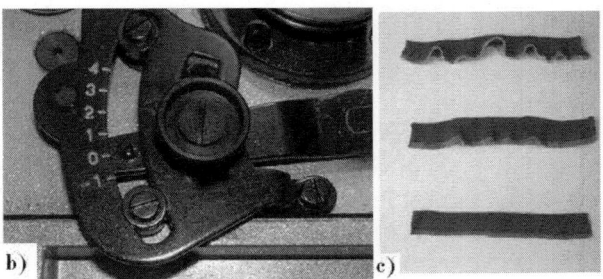

Figure 2.53 (a) Overlock sewing machine differential feed adjustment screw, (b) Mark for identification and (c) Produced samples.

2.7.4.3 Adjustment of stitch length and width

- To adjust the stitch density or stitch length, pulley wheel should be turned by pressing the push button as shown in Figure 2.54(a). At some point, it will be felt that the press button goes inside and becomes fully settled.

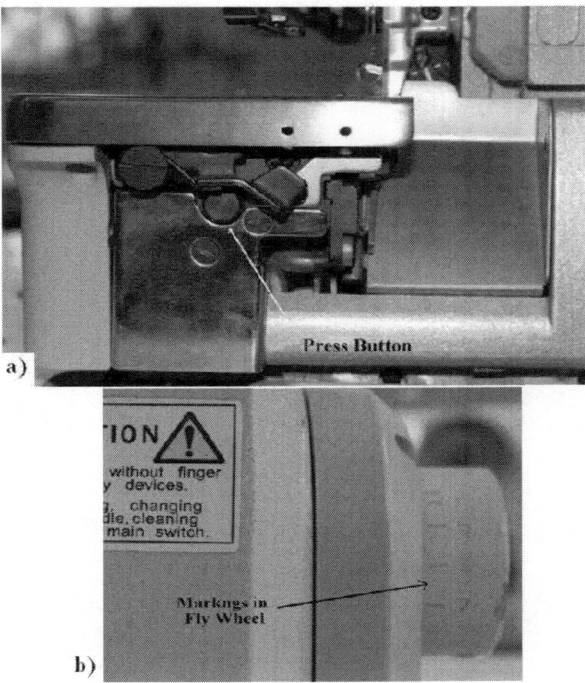

Figure 2.54 (a) Overlock machine stitch density regulator button, (b) Mark for identification of stitch density in fly wheel.

- The process of pressing the press button should be continued by turning the pulley and the higher number marked in the flywheel should be matched with the arrow mark provided in the machine (Figure 2.54b).

- This provides stitch with high density. If arrow mark is matched with the lower number, a stitch with less density per unit area is obtained. The stitch width of an overlock machine is determined by two or three factors. On some machines the stitch finger or the cutting knife can be moved by turning a dial or loosening a screw as shown in Figure 2.55. Or it can be carried out by changing the throat plate to one that has a wider or narrower stitch finger.

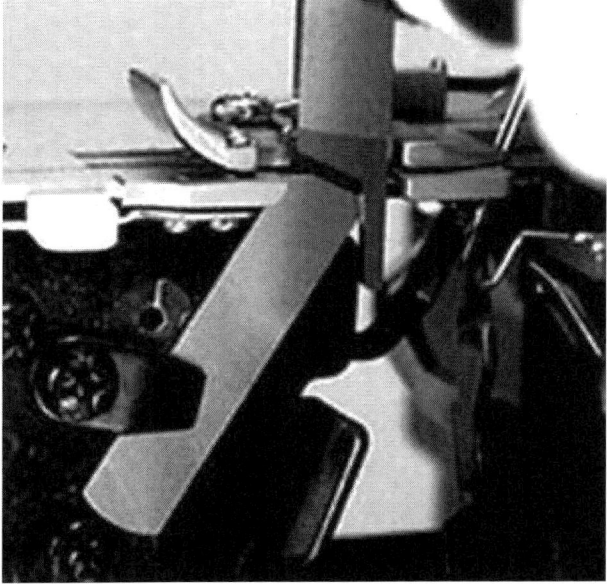

Figure 2.55   Overlock sewing machine cutting knife.

2.7.4.4 Adjusting the lift of the presser foot

- In the back side of the overlock machine, the adjustment setting is available.

- The machine pulley should be turned until the tooth of the bottom feed dog is below the top surface of the needle plate during the travel.

- In Figure 2.56a, the nut 2 should be loosened and the presser foot lever 3 should be lowered till it touches the screw 1.

- Now the height of the presser foot is adjusted by turning the screw 1 as required so that its bottom surface touches the top surface of the needle plate.

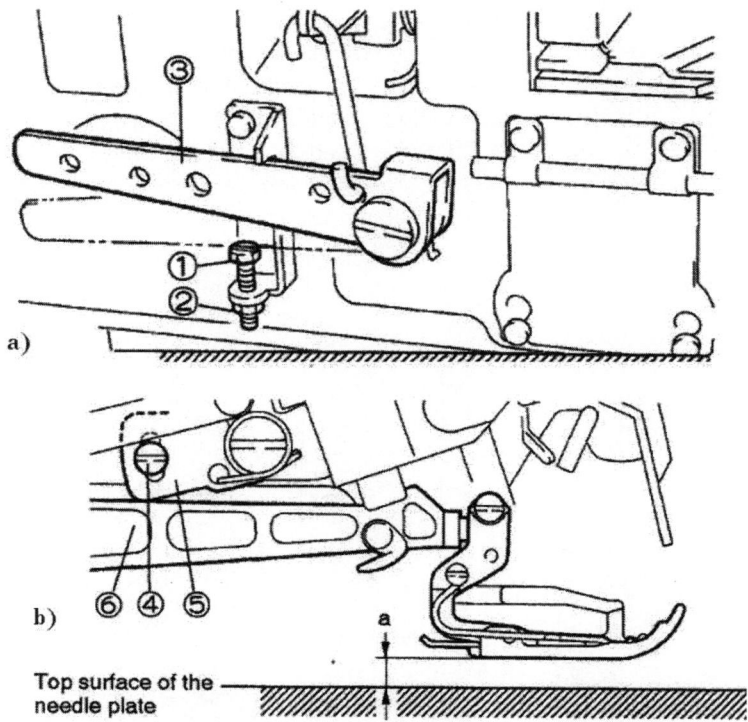

Figure 2.56   (a) Overlock machine pressure foot lift adjustment at back, (b)
Adjustment in the front side.

- The presser foot lever 3 should be replaced and the nut 2 should be
  tightened. Screw 1 should be securely fastened.

In the front side of Figure 2.56b, presser foot should be raised and the screw
4 should be loosened. Then the height (a) should be adjusted by adjusting the
stopper 5 and securely tightening the screw 4.

2.7.4.5 Timing and settings between looper and needle

Top looper setting

- The needle is kept at the top most position. The looper is manually
  moved and placed in front of the needle in such a way that the looper
  tip is projecting outside the needle.

- The distance between needle and looper tip is 4 mm approximately (a)
  in Figure 2.58.

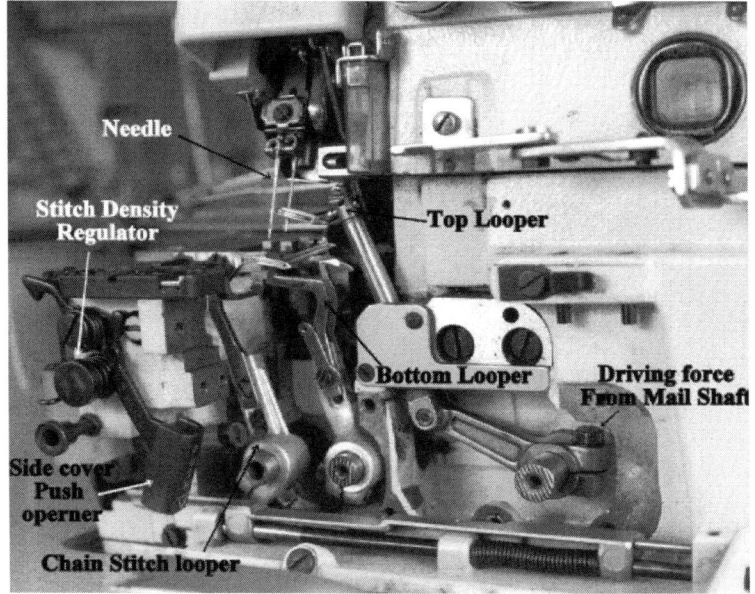

Figure 2.57 Overlock machine looper and stitch formation device.

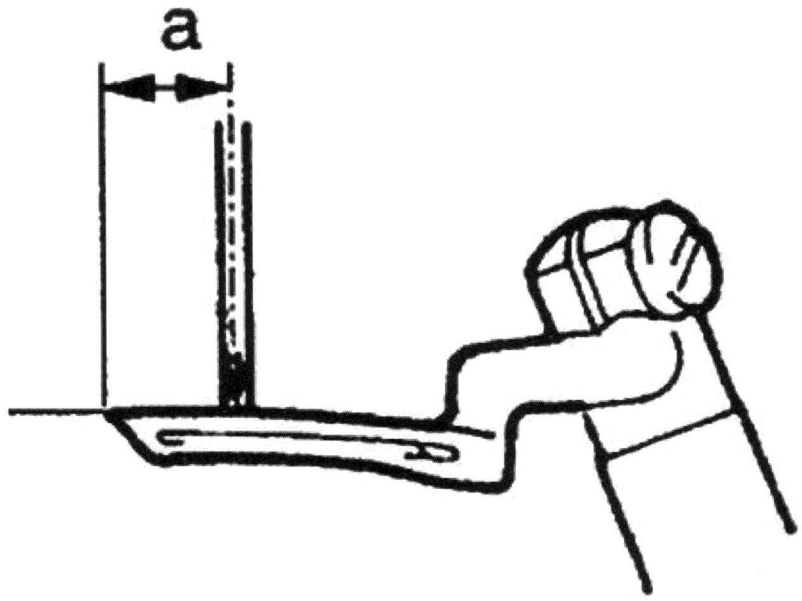

Figure 2.58 Top looper setting point.

Bottom looper setting

- The needle is kept at the bottom most position.

- The looper is moved near to the outside needle.

- The space between needle and looper tip is kept at 3 mm approximately (b) as in Figure 2.59.

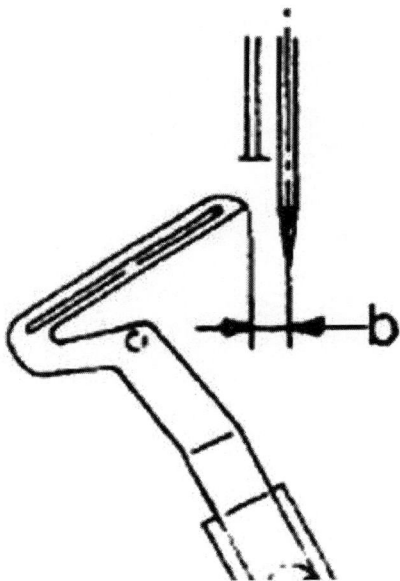

Figure 2.59   Bottom looper setting point.

The adjacent clearance setting between the top and bottom looper should be 0.2 mm (c) and 0.5 mm in front and back, respectively, as given in Figure 2.60.

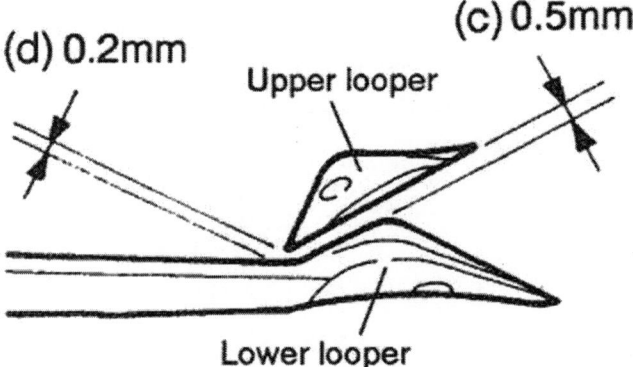

Figure 2.60   Setting point between top and bottom loopers.

Chains stitch looper

- The needle is kept at the bottom most position

- The looper is brought to the side of front needle (chain stitch needle)

- The space between the needle and looper is kept at 2 mm approximately (a) as provided in Figure 2.61.

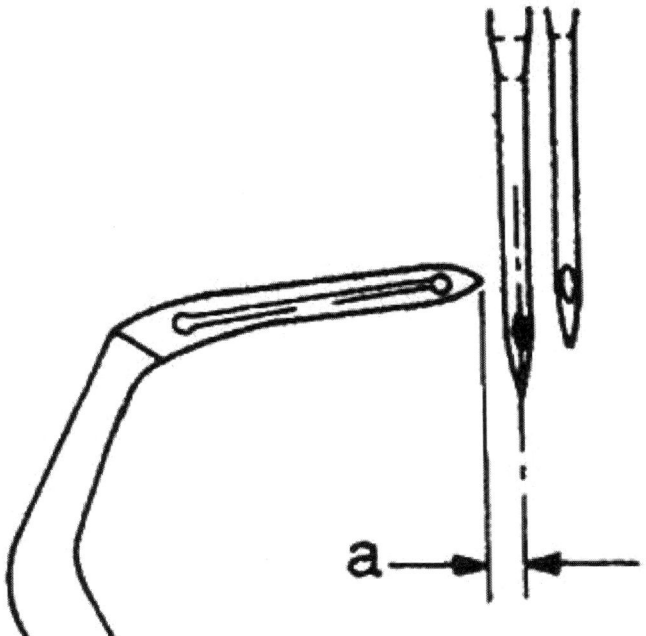

Figure 2.61   Setting point between chain stitch looper and needle.

2.7.5 Chain stitch machine timing diagram

- At 0°, the needle bar will be in throat plate or base position

- The upper looper/first looper will move forward (during the movement from 0° to 90°) and the lower looper will move backward

- At 90° needle bar will be in higher most position. The first/upper loop will be in throat plate/base position. Second looper/lower looper will be in backward position

- At 180° the first looper will return back, the first stitch/loop will be formed and the needle bar will start moving further down

- At 270° first looper will be in backward position, lower looper will move towards (during the change from 180°–270°) and at 270°, it will form the second loop/stitch.

- At 360° after the stitch formation, lower loop will move back and needle bar will rise to its base/throat plate position and the next cycle will begin this was given in Figure 2.62.

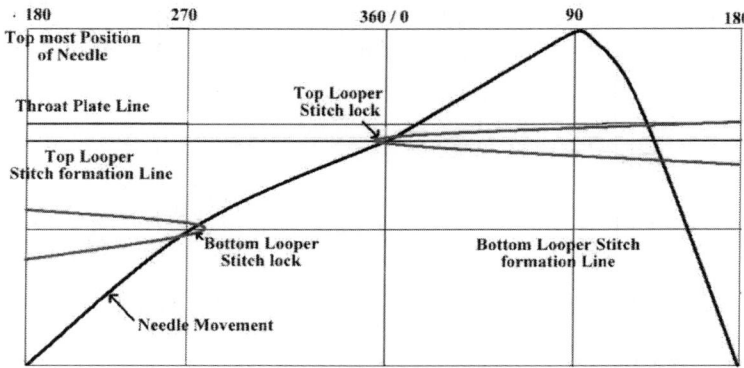

Figure 2.62   Chain stitch / Overlock machine stitch formation - Timing Diagram.

2.8   Flatlock sewing machine

The flatlock stitch is used decoratively to achieve the look of applied trim on a garment, craft or home decorator projects. Special decorative threads are used for a textured and dramatic embellishment effect. The flatlock stitch can also be used to serge non-bulky seams. The flatlock stitch is sewn with one needle and the upper and lower loopers. Both sides of the flatlocking stitch are considered decorative. The majority of the flatlock machine parts remain common as discussed in overlock machine. However, there are changes in the machine settings which are discussed in detail further. The important parts of flat lock machine are shown in Figure 2.63.

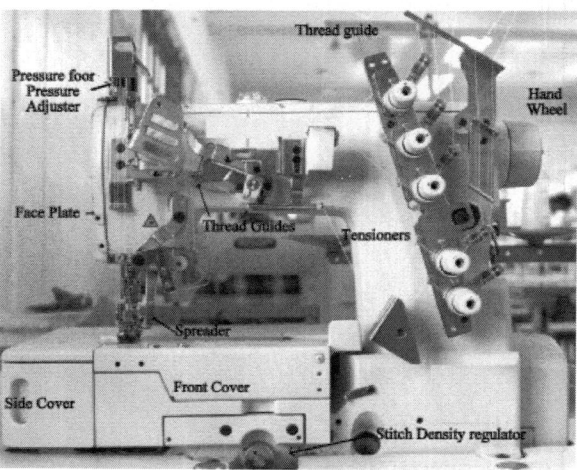

Figure 2.63   Parts of Flat lock sewing machine.

2.8.1 Working mechanism of flatlock machine

- The flatlock sewing machine gets drive from the electric motor which is mounted in sewing table. The main parts of the sewing machine are flywheel, main shaft, needle bar, loopers, feed dog and presser foot.

- The motion from the main shaft is directly connected with the needle bar mechanism. A link from the main shaft is taken to spreader mechanism with a cam. The cam rotation is fixed as half revolution per every full rotation of the main shaft. So that, the spreader movement is controlled (Figure 2.64).

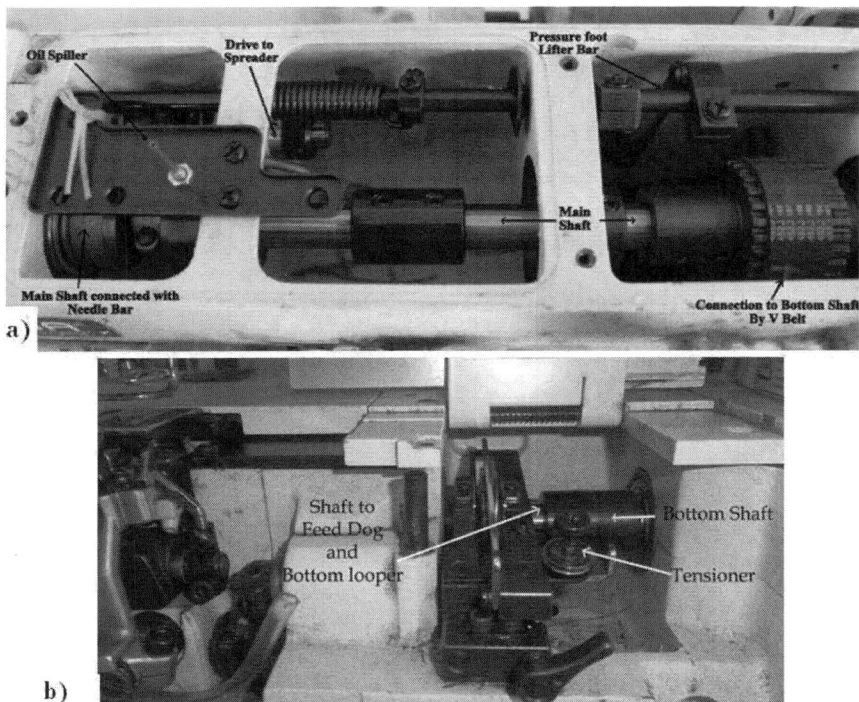

Figure 2.64 Flat lock sewing machine – Top and bottom main shaft mechanism.

The knee presser foot lifter mechanism is also connected to presser foot bar in the top side. By a 'V' belt, the drive is transferred to the bottom of the machine and the bottom shaft is directly connected to the feeding mechanism. The bottom shaft also controls the movement of the bottom looper. A cam attached with the bottom acts as a tensioner device which controls the sewing thread supply to the bottom looper.

2.8.2 Height of the needle bar

Figure 2.65 shows the side view of the machine, with needle bar and presser foot mechanism. When the needle bar is lifted to highest position, the standard setting between the left needle point and the needle plate are as follows:

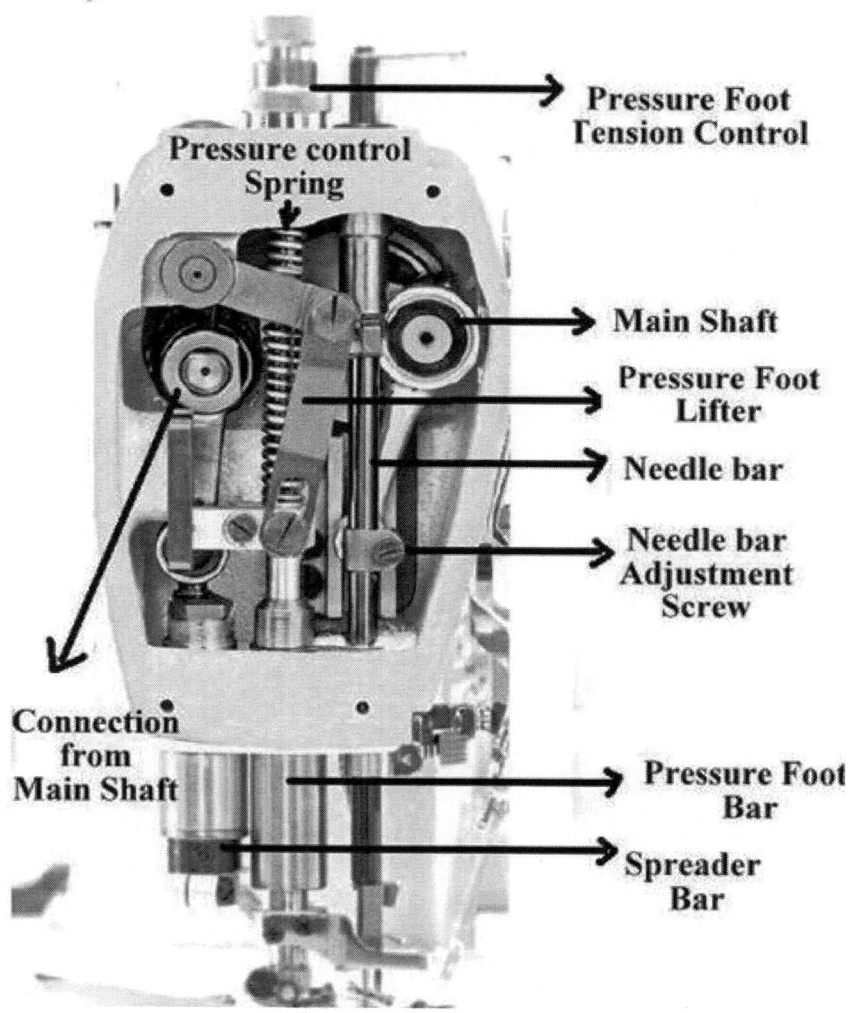

Figure 2.65   Needle bar set up for flatlock machine.

- For 2-needle machine – left needle height – 9.3 mm (needle gauge – 4.00 mm)

- For 3-needle machine – left needle height – 8.5 mm (N.G – 5.6); 8.1 mm (N.G – 6.4 mm)

To adjust the needle bar height, the face plate is removed and the needle bar and cam connecter screw are loosened, the height is adjusted as required and the screw is securely tightened.

2.8.2.1 Presser foot – lift height adjustment and pressure adjustment

The lift height of the presser foot refers to the distance between the bottom of the presser foot and the top of the needle plate when the presser foot is in its highest position.

When the presser foot (A) is raised, the distance B in Figure 2.66 can be adjusted from 3.5 mm to 7.0 mm based on the requirement. For this purpose, the machine pulley is turned until the feed dogs are at their lowest position of travel.

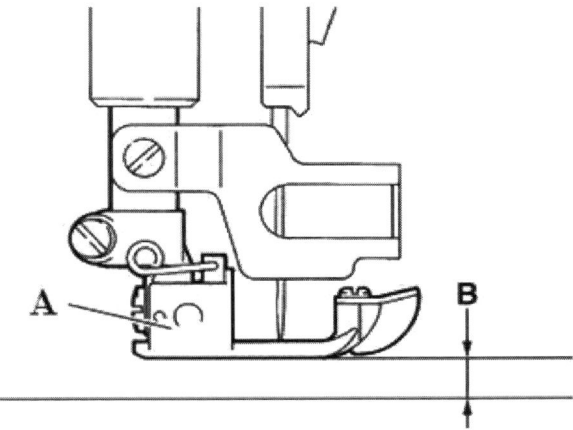

Figure 2.66   Pressure foot lift set up for flatlock machine.

In Figure 2.67, the presser foot lift is adjusted by lowering the lever Z. (Presser foot should not contact the spreader). By adjusting screw C and D, the lever Z can be adjusted. Then the screw C is tightened at the back side of the flat lock machine.

2.8.2.2   Pressure adjustment

By moving the presser foot pressure regulator (a) the presser foot pressure can be increased or decreased on the fabric in Figure 2.68. The provision shown in Figure 2.68b is at the back side of the machine which is called presser foot micro-lifter mechanism that is specially designed for stitching elastic kind of material. For stitching elastic or light weight material, the presser foot is kept slightly raised during sewing process by adjusting the screw. This helps problems like degree of slippage, warpage and damage of the light weighted material effectively.

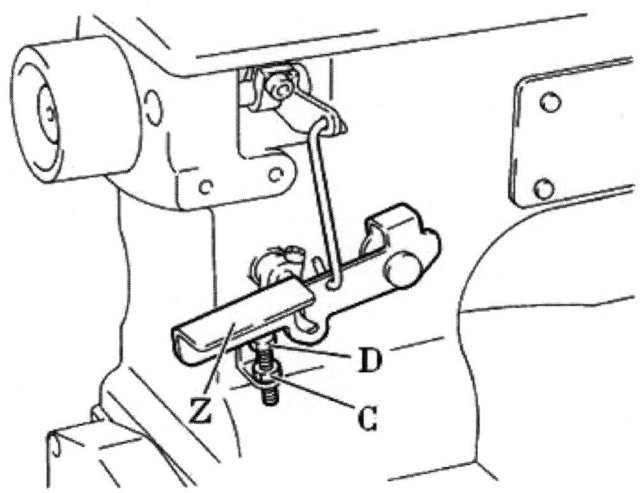

Figure 2.67   Presser foot lift adjustment for flatlock machine.

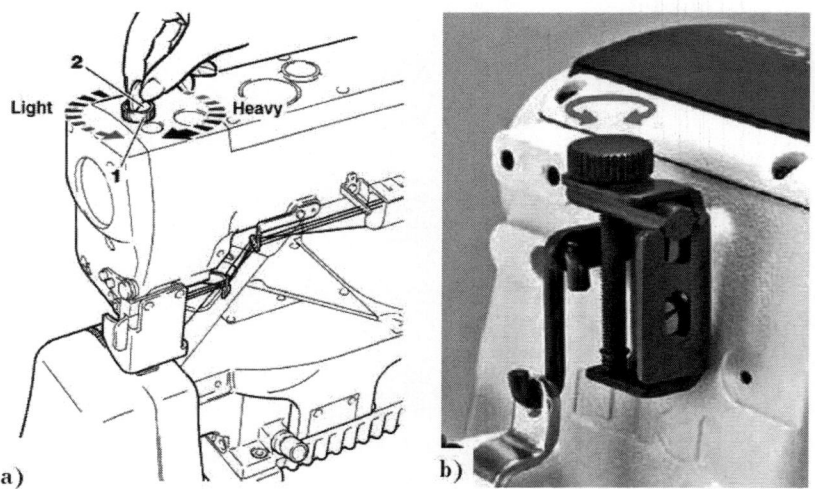

Figure 2.68   (a) Presser foot pressure set up for flatlock machine (b) Micro
level adjustments.

2.8.3 Adjusting the stitch length

Method 1

In modern machineries, the stitch density regulation adjuster is provided
in the front side of the machine. By simply turning it in clockwise or anti-
clockwise direction, the stitch density can be controlled (see Figure 2.69).

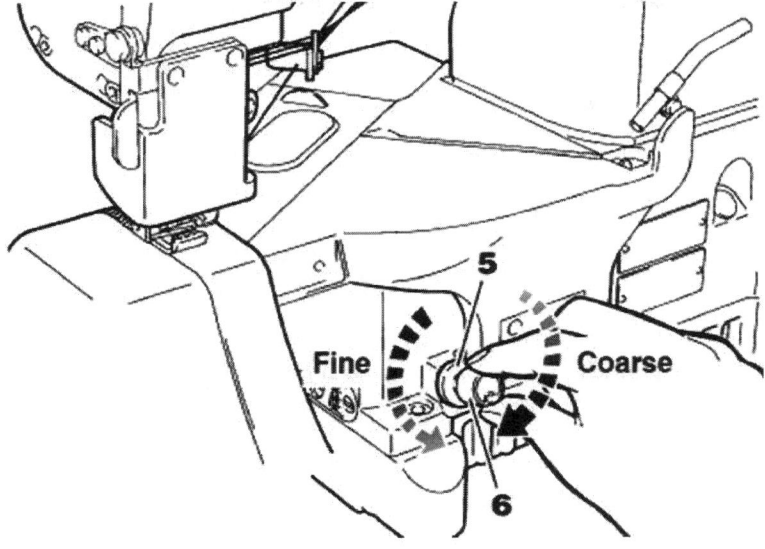

Figure 2.69   Stitch density adjustment (Method 1).

Method 2

Push button is pressed first and held; simultaneously the hand wheel is turned to seek the position at which the push button goes further inside. With the push button kept pressed, the hand wheel is turned and the required scale is set matching with the mark. Then the push button is released to fix the stitch length (Figure 2.70).

Stitch Density
Regulator

Differential Feed
Regulator

Figure 2.70   Stitch density adjustment (Method 2).

2.8.4 Adjusting the differential feed

- Similar to the overlock machine, the flatlock machine is also provided with 2 separate feed dog set up. The adjustment in the feed dog movement will alter the feeding ratio between the front and rear feed dog as in Figure 2.71a.

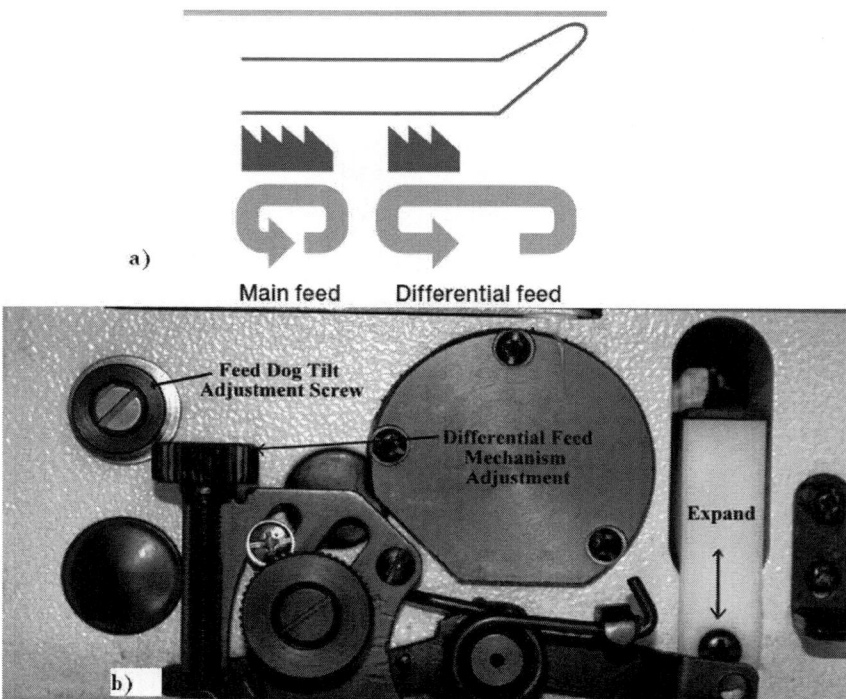

Figure 2.71 Differential feed adjustments.

- If the distance moved by the differential feed dog is larger than that of main feed dog, then the movement is called as normal differential.

- When the reciprocating distance of the differential feed dog is smaller than that of main feed, then the feed system is called reverse differential.

- The differential feed ratio is the ratio of the distance travelled by the main feed dog to the differential feed dog. If the main feed dog travels 3 mm and the differential feed dog travels by 6 mm, the differential feed ratio is 3:6, i.e., 1:2.

2.8.4.1 Feed dog position (height and tilt)

When the feed dogs are at their highest point of travel, there should be a distance of 0.8 mm–1.2 mm from the top surface of the needle plate to extended line 'A' in Figure 2.72, from the tips of main (1) and differential feed dog (2).

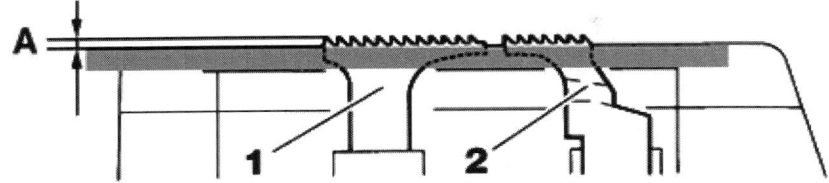

Figure 2.72   Feed dog height adjustment for flatlock.

To adjust the feed dog height, the screws 3 and 4 (in Figure 2.73) can be loosened and the feed dog can be adjusted.

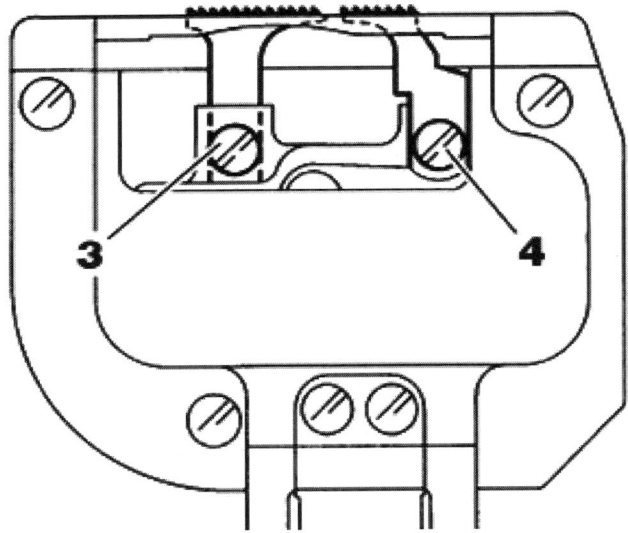

Figure 2.73   Feed dog inclination adjustment for flatlock.

The position of the feed dog can be adjusted based on the requirement like backward tilt and forward tilt. In the case of backward tilt, the front side of the feed dog is raised by adjusting the screw counter clock wise. The front side of the feed dog is lowered down in the case of the forward tilt of the feed dog. This can be achieved by adjusting the screw in the clockwise direction (see Figure 2.74).

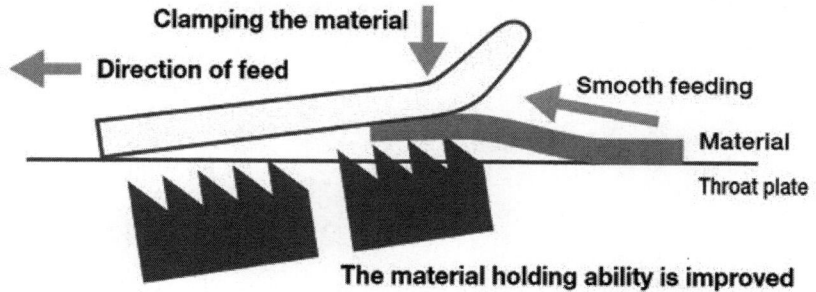

Figure 2.74   Feed dog inclination purpose.

2.8.5 Needle to looper timing and setting

During the needle insertion, the needle is inserted in to the groove to the maximum level.

1. Any fixation not up to the maximum insertion like 2 will affect the stitch formation quality. Always the needle is fixed by facing the scarf to the back side of the machine.

The distance between needle and the machine throat plate has to be checked every time. To adjust the distance 'a' from the point of left needle to the top surface of the needle plate, the needle is kept at the top most position (Figure 2.75). The distance varies from 7 mm to 8.2 mm based on the machine type.

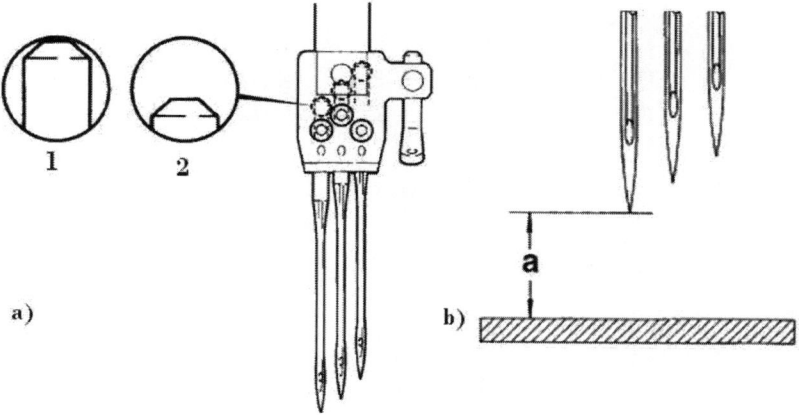

Figure 2.75   (a) Flatlock machine – Needle fixation (b) Needle height.

Figure 2.76 shows, basic flat lock machine parts in the feeding system. The needle is kept at the bottom position, the looper screw is loosened and manually the looper is adjusted to a position where, the distance between the looper tip and the inner most needle (1) of the flatlock machine is approximately 3.8 mm (A) by adjusting the screw at the bottom of the looper (Figure 2.77).

Figure 2.76   Flat lock machine looper set up.

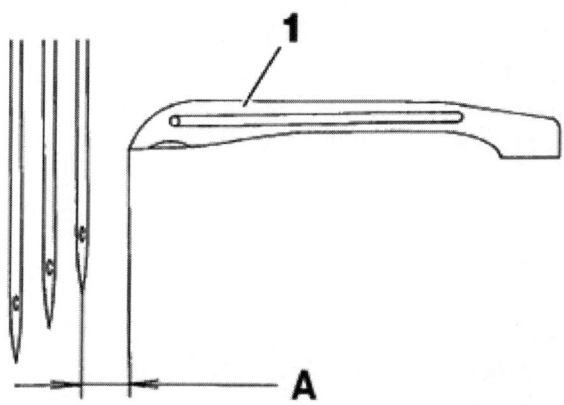

Figure 2.77   Top looper and needle setting point.

When the point of the looper has reached the centre of the left needle from
the extreme right end of its travel, there should be 0.05 mm to 0.1 mm front to
back clearance between the left needle and the point of the looper. During this
travel, after passing the centre needle to the extreme right end, the right needle
slightly contacts the point of the looper (Figure 2.78).

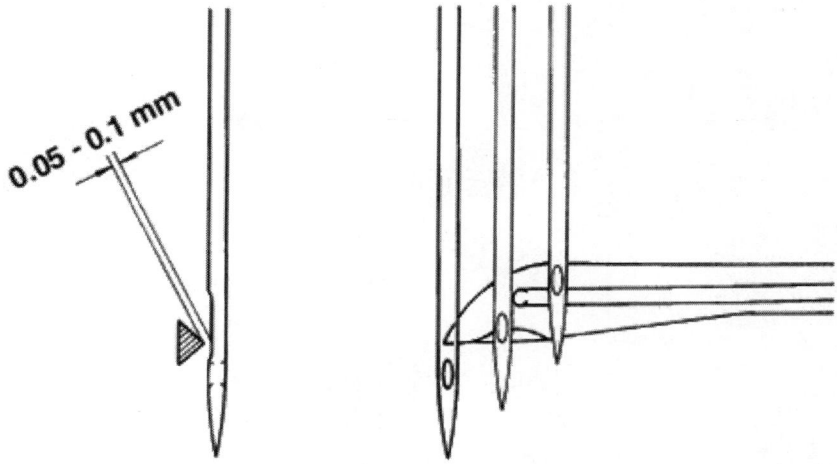

Figure 2.78   Looper and needle back to front set up.

The needle is allowed to lower and move the bottom looper to move to
the extreme end of the outside needle. The setting is kept as such, where the
distance between the outside needle and the tip of the bottom looper is 1.5 mm
approximately as in Figure 2.79.

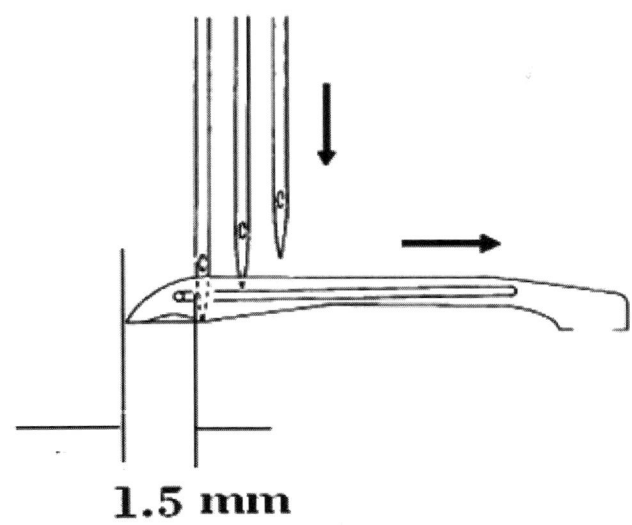

1.5 mm

Figure 2.79 Needle to Looper setting and the right extreme.

During this setting process, while the needle moves from inner most needle to outer most needle, it should be aligned in such a way that, for all the needles, the eye of the looper should match with the eye of each needle. This helps the looper tip to pick the thread from the needle scarf.

In this process, during needles downward movement, the looper passes at the back side of the needle. Once the needle starts to return from the downward most position, the looper starts to move in front of the needle and completes its elliptical movement. All these processes around needle take place below the feed dog. This mechanism helps to lock the needle stitch after every stitch formation.

2.8.6 Top looper or spreader settings

2.8.6.1 To adjust the spreader height

To adjust the distance A from top surface of the needle plate to the bottom surface of the spreader (1) in Figure 2.80, the screw 2 is loosened and the height is adjusted from 7.8 mm to 9.0 mm based on machine type and then the screw is tightened.

To adjust the spreader front to back, when the point B (in Figure 2.81) on the spreader comes close to the left needle while the spreader (1), is being moved to the left from the extreme right end of its travel, the distance of

0.5 mm is kept from point B to the left needle. This can be done by adjusting the screw 2 and moving the 1 front and back.

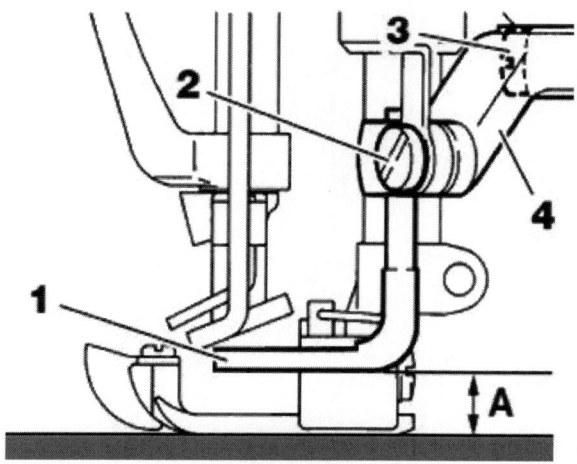

Figure 2.80 Needle plate to Looper setting.

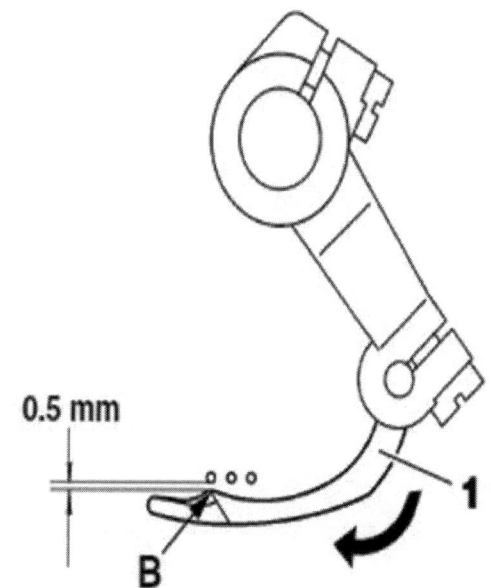

Figure 2.81 Needle back to Looper front distance setting.

For left to right adjustment, when the spreader is at extreme left end of its travel, there should be a distance of 0.5 mm (as in Figure 2.82) from the centre line of the left needle point B on the spreader. Adjustment is made by loosening the screw 3.

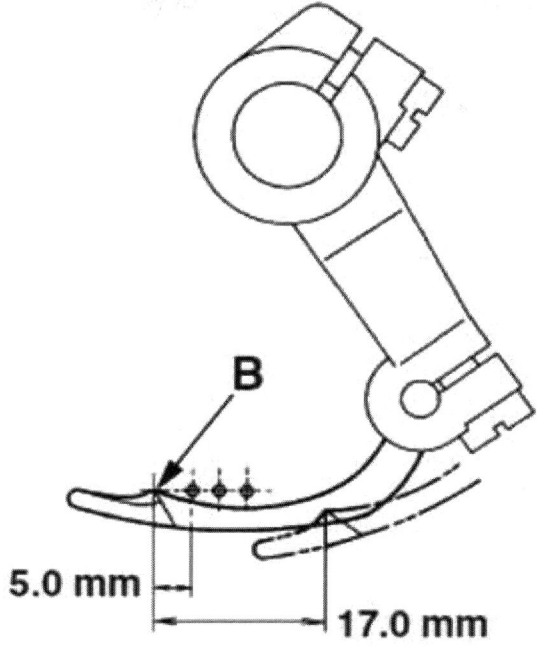

Figure 2.82   Needle to Looper point setting.

2.9   Summary

The chapter has provided a completed view of the working and principle behind various sewing machines. The important and the most highly used basic sewing machines – Single needle lockstitch sewing machine (SNLS), double needle lockstitch sewing machine (DNLS), overlock and flatlock machine purpose, construction, working and various setting points are discussed in detail in the chapter. The classification of sewing machines is provided with a brief introduction on the evolution of the sewing machinery. The highlight of this chapter is the detailed explanation on the various setting points of the basic sewing machines. Clear illustrations are provided for easy understanding of the working and setting points with step-by-step explanation on how to keep the settings.

References

1.   Carr, Harold and Latham, Barbara, *The Technology of Clothing Manufacture* – 4th edition, Blackwell Publishing, Oxford, UK, 2000.

2. Solinger, Jacob, *Apparel Manufacturing Handbook-Analysis, Principles and Practice*, Columbia Boblin Media Corp., 1988.

3. Kratzer, Constance, *Sewing Techniques with an Overlock Machine*, College of Agriculture and Home Economics. www.cahe.nmsu.edu.

4. http://www.siruba.co.uk/partslists.htm.

5. http://texeducation.wordpress.com/2013/12/24/principle-of-lock-stitch-formation/.

6. http://home.howstuffworks.com/sewing-machine.htm.

7. http://www.sewalot.com/sewing_machine_history.htm.

8. www.wikipedia.com.

9. Kerr, Douglas A, , *Stitch Formation in Rotary Hook Sewing Machine*, 3, pp. 1–11, 2008.

3

Feed mechanism and lubrication systems

The essential part of the sewing machine is feeding mechanism. This chapter describes the different types of feeding mechanism and their functional requirements on sewing machine. The elements of feeding section and their importance are discussed. The working of feeding mechanism is explained with clear illustration. The second part of the chapter discusses about the different lubrication systems used in the sewing machine and their working.

> **Key words**: Feeding mechanism, Classification, Applications, Lubrication systems.

3.1   Elements of feed mechanism

It is one of the controls available on sewing machine. It controls the length of the stitching.

The feeding mechanism controls the amount of fabric fed by feed dog during every sewing cycle of needle. The three sewing machine parts, which together constitute the drop feed mechanism are presser foot, throat plate and feed dog. They are also known as the basic elements of feed mechanism.

3.1.1 Presser foot

- The presser foot can be raised and lowered with a small lever at the back of, or beside, the needle.

- When the presser foot is up, it permits the movement of the fabric freely.

- When the presser foot is down, it presses the cloth against a base plate.

- The base plate has a couple of textured moving parts (the feed dogs) that keep the material moving past the needle at an even rate. Figure 3.1 represents the basic elements of feeding mechanism.

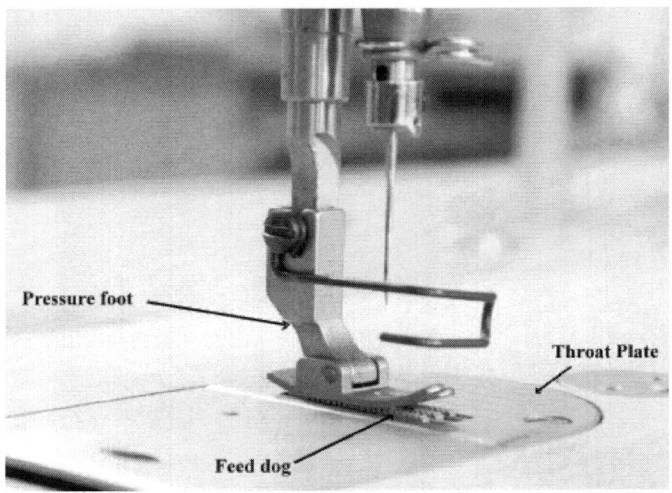

Figure 3.1   Parts of feeding mechanism.

3.1.2 Feed dog

- During the stitching process, after every successive stitch, predetermined amount of fabric is moved by feed dog. In normal single needle lockstitch machine, typically a set of feed dog with two or three rows will be present.

- These feed dogs typically resemble short, thin metal bars, crosscut with diagonal furrows, which move back and forth in grooves which are slightly larger than the bars. The feed dog makes four motions.

  - Forward – down – backwards – up: These four steps of the feed dog helps the machine to pull the fabric while stitching. The 'feed dogs' are in contact with the material on the forward stroke, and are pulled down below the main plate on the backward stroke by the sewing machine's mechanism.

- The feed dog consists of a toothed surface which rises through the openings in the throat plate and engages the under surface of the fabric. It takes the fabric along towards the back of the machine and then drops it below the throat plate before commencing the whole cycle again.

- Once the cycle is completed, the feed dog drops below the throat plate. In this time, this plate supports the fabric from falling away and so that it loses contact with the feed dog and is not carried back with it.

- The up and down direction movement of the needle in the sewing process must be synchronized accurately with the four step elliptical motion of the feed dog. This helps the feed dog to pull the fabric only during the time when the fabric is not in contact with the needle. It can be observed that, the fabric movement during the sewing process happens as series of discrete steps, even though it appears to be a continuous motion.

- In a sewing machine, the number of feed dog, position, length and nature of the toothed surface of the feed dogs can be varied at any situation based on the application requirement. The rationale behind this is, when a single row or short length of the feed dog provides very less surface contact and reduced friction between the material and feed god, it creates the problem of fabric slippage during sewing either of the sides instead of passing in a straight line.

- In an over edge machine, the feed dog is usually mainly to the left of the needle drop point, because it trims and sews the fabric to the right of the needle and also there is a chaining-off finger on the throat plate over which the stitch is formed. The teeth on the surface of the feed dog can be of different types and sizes but they are generally slanted slightly towards the direction of feeding. For sewing of light to medium weight fabrics, a tooth pitch (distance from peak to peak) of 1.3–1.6 mm is normal. On very lightweight fabrics, sagging can occur between the teeth and pucker can appear after sewing as a result. Fine-toothed feed dogs with a pitch of only 1.0–1.25 mm can be used to prevent this. On heavyweight fabrics, a certain amount of sagging is required for satisfactory feeding in order to keep both plies together. In this case, coarser feed dogs of 2.5 mm tooth pitch may be needed. Based on the fabric type and requirement, rubber-coated feed dog with no sharp teeth can also be used for very delicate materials like soft leather and soft fabric.

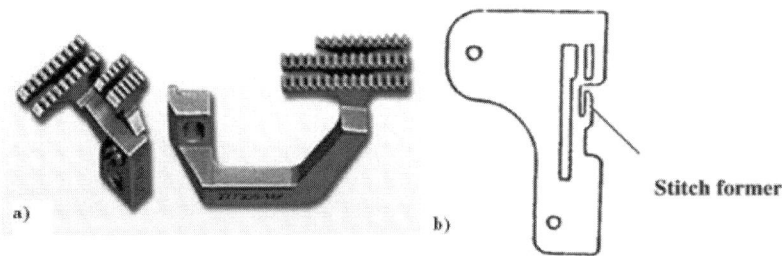

Stitch former

a)

b)

Figure 3.2   (a) Feed dog and (b) Throat plate.

- If the shape of the feed dog is sharp at the top end, it may damage the material and cause thread breakage or chain off thread break. To avoid this, the feed dog may be grinded with grindstone on the surface slightly.

• If the width of the flat surface at the top of tooth is 0.1 mm or more, then decrease the feed force or else there is a chance for uneven pitch or material slippage to occur (Figure 3.3).

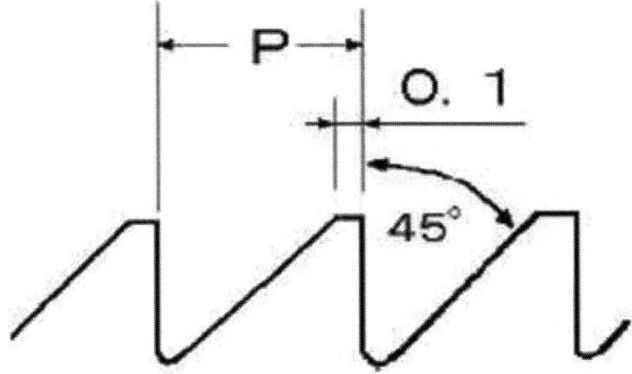

Figure 3.3   Pitch length of the feed dog.

3.1.3 Throat plate

• The main function of the throat plate is to provide a smooth, flat surface over which the fabric passes as successive stitches are formed. It has one or more slots in it which matches the sections of the feed dog.

• It has a hole through which the needle passes as it goes up and down. The needle hole should be only about 30% larger than the size of the needle. If the hole is too large, the fabric may be pushed into the hole through each penetration of the needle. This problem is known as 'FLAGGING' and causes missed stitches and yarn breaks.

• Fabric flagging is a machine-related issue; the throat plate aperture enlarges due to wear and tear, or while sewing the needle pushes the fabric through the aperture before penetrating the fabric. This can also happen when the needle size (thickness) is changed and if the throat plate is not changed accordingly as shown in Figure 3.4.

• Throat plates must be changed at regular intervals after checking for wear and tear.

• Throat plates must be changed in accordance with the needle size even if there are no signs of wear and tear.

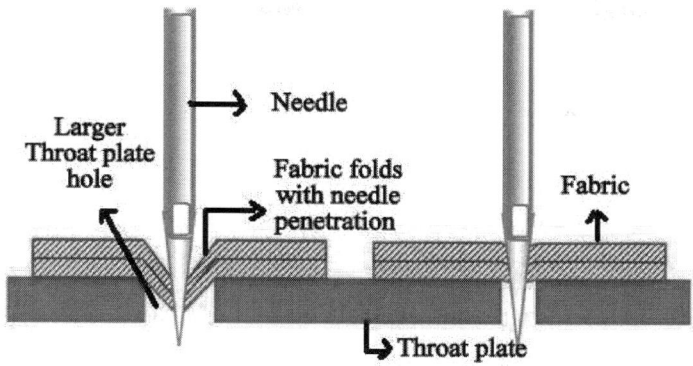

Figure 3.4 Throat plate damage.

3.2 Types of feed mechanism

Based on the end use and application, the feed mechanism can be classified into following types:

- Manual feed or free motion or freehand or darning feed
- Drop feed
- Differential feed
- Needle feed
- Compound feed
- Unison feed
- Puller feed
- Cup feed
- Wheel feed
- Clamp feed

3.2.1 Manual feed or free motion or freehand or darning feed

- Manual feeding mechanism is a simple one where the operator moves the work material under the needle. It is a free-hand motion without any feed dogs and feeding mechanism.

- Machine may have a vertical motion foot that clamps the goods before the needle enters the material and releases to allow the operator to manipulate the goods between each stitch, for example, darning, embroidery, freehand quilting, etc.

3.2.2 Drop feed

- Drop feed mechanism is the most common feed system, also called four-motion drop feed, it incorporates a feed dog as shown in Figure 3.5.

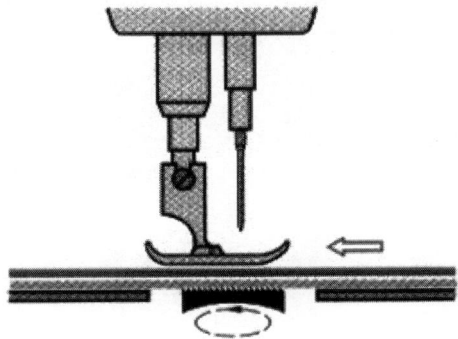

Figure 3.5   Drop feed mechanism.

- In this drop feed mechanism, the feed dog below the needle plate starts and rises up through the provision in the plate, compresses the fabric against the presser foot to advance the fabric one stitch, then drops below the plate to return to the original position.

- In this mechanism, the movement of the fabric totally depends on the feed dog movement because the friction between the teeth of the feed dog and fabric plays a vital role in the fabric movement. Thus, successful feeding of more than one ply is dependent on friction that enables holding the plies together.

3.2.2.1 Limitations of drop feed

The friction between the bottom ply and the feed dog is normally greater than the friction between intervening plies. This makes the lower ply to move as expected but the only control for the upper ply and middle layers is the friction between the fabrics. In that case, if the fabrics are off different fibre materials, then there will be a different friction level that leads to slipping of fabric layers as shown in Figure 3.6(a). The problem is known as inter-ply shift or differential feeding pucker or just feeding pucker.

In sewing a hem, twisting may occur between the layers of material and the problem is known as 'roping'. When pressed, the seam may show a curve because of the excess of the fabric along one side as shown in Figure 3.6(b).

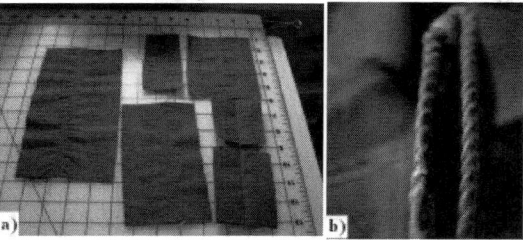

Figure 3.6 Problems caused by drop feed mechanism.

3.2.3 Differential feed

The differential feeding mechanism works as same like drop feed mechanism but the only difference is that this mechanism has a pair of individual feed dogs under the needle plate. These feed dogs can work individually.

- Out of the two feed dogs, the front (main) feeder and rear feeder can be set to move the same or different distances as shown in Figure 3.7.

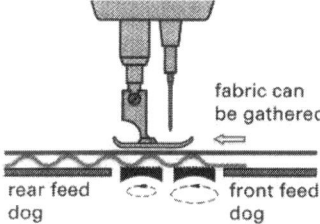

Figure 3.7 Differential feeding mechanism.

- During the fabric feeding process, if the rear feeder sets to move faster than the front one then the fabric being sewed gets stretched because of the higher speed of the rear feed dog as shown in Figure 3.8(a).

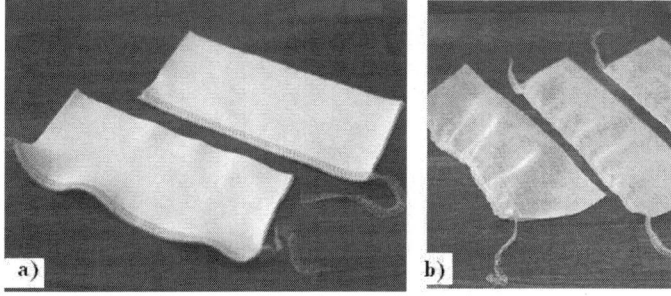

Figure 3.8 Fabric stretching (a) and gathering effects (b) produced by differential feed mechanism.

- In the contrary, if the front feeder moves faster than the rear feed dog, higher amount of fabric is fed to the rear feeder more than its capacity. This makes the fabric to gather as shown in Figure 3.8(b) during stitching process and forms the shirring effect on the sewn material. Differential feed is available on chain stitch, overedge, safety stitch and also on lockstitch machines.

- The movement of each section is similar to the movement of the whole feed dog in the drop feed system, but the stroke or movement of each part can be adjusted separately or differentially.

3.2.4 Top feed

Top feeding mechanism is a special feeding mechanism, where the top presser foot connection is split into two separate sections as mentioned in Figure 3.9.

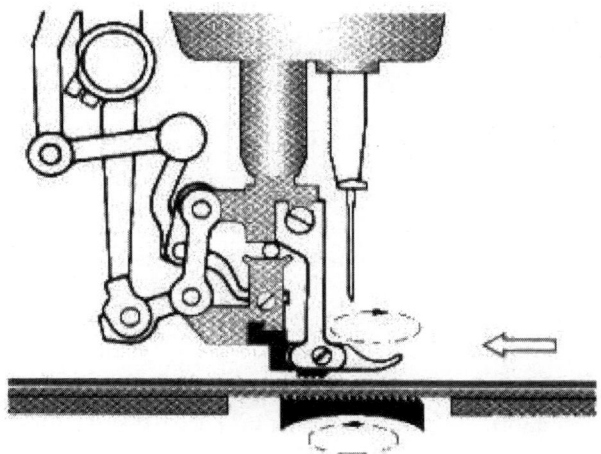

Figure 3.9   Top feed mechanism.

- This extra section of the presser foot (2), as shown in Figure 3.10, provides the positive control on the top ply of the sewing item by pressing the material against the needle plate during the fabric movement time also, and this presser foot moves back along with the fabric.

- This helps the fabric to be in position during the movement without slipping from the actual requirements when the speed of the presser foot matches with the bottom feed dog movement. This positive control of the presser foot also helps to alter the movement of the top ply. Thus as like differential feed mechanism, the gathering or shirring can be created in the top ply. But in the case of differential feed mechanism it is possible only on the bottom ply.

- The general arrangement of such top feed systems is that the presser foot is in two sections, one holding the fabric in position, while the needle forms the stitch, and the other having teeth on the lower side and moving or 'walking' in such a way that the top ply is taken along positively while the needle is out of the material. All feeding systems with two presser feet can be described using the name 'walking foot', shown in Figure 3.10 as (2).

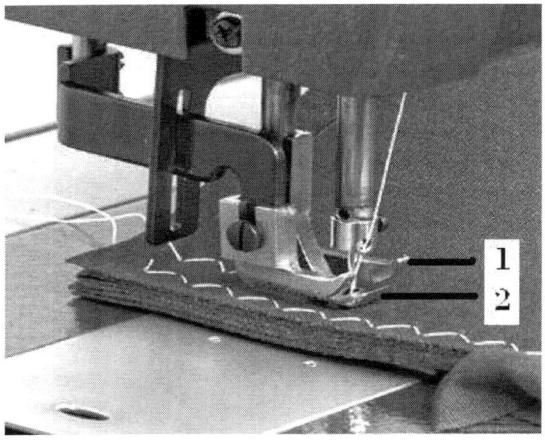

Figure 3.10   Working of top feed mechanism.

- In applications, the variable top feed can be combined with a differential bottom feed to create fancy effects either on the top or on the bottom ply and for achieving shift-free sewing. Such a combination of feeding systems with both the top feed and the bottom feed being independently variable is shown in Figure 3.11.

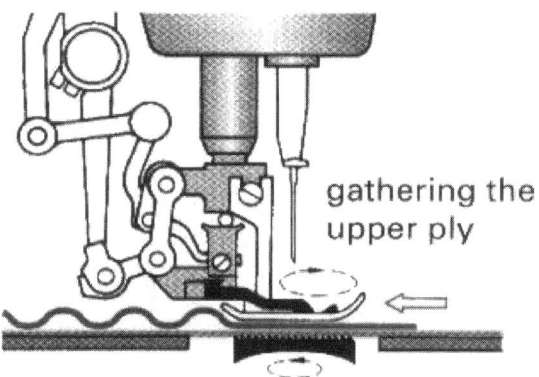

Figure 3.11   Positive control on the top ply.

3.2.5 Needle feed

Needle feed is the type of feed mechanism in which the needle itself moves forward and backward during the sewing process as shown in Figure 3.12. The machine utilizes the feed dog below the plate that rises up through the plate, compresses the fabric against the presser foot in conjunction with the sewing needle which drops through the fabric and then both move one stitch to advance the fabric. Then they separate and return to the original position for the next stitch.

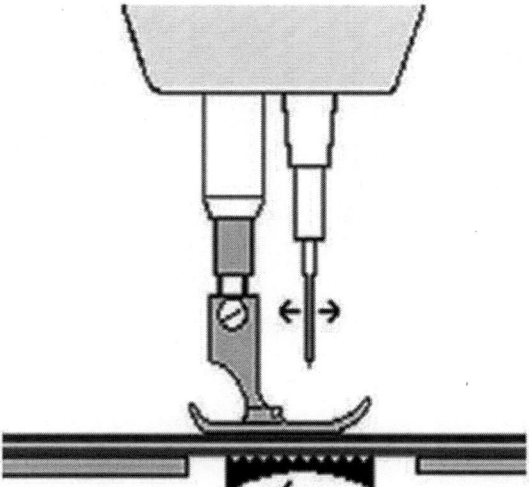

Figure 3.12   Needle feed mechanism.

- During the stitching process, the needle enters the fabric, when the feed dog moves the fabric back, the needle also moves back with it for the completion of one stitch and then rises up and forward again to begin the next stitch. Thus, the needle is in the fabric while feeding is taking place and the plies of fabric are held together.

- In Figure 3.13(a) it can be noticed that the needle is moving down and the needle bar is in front most position. In this state, the distance between the needle bar and presser foot bar is very large as indicated. In Figure 3.13(b) the needle starts to come down, and in Figure 3.13(c) the needle is at down most position inside the fabric plies. It moves a step back as fabric moves after sewing, the presser foot raises and allows the action. It can be seen in the figure where, the distance between needle bar and presser foot bar is reduced and the needle is in slanting position.

- In Figure 3.13(d) from the backward most position, the needle starts moving to its original top most position. In this mechanism needle enters into the fabric at a leading angle and leaves the fabric in a trailing angle. Hence, sometimes there are possibilities for extended needle holes in sewn fabric as a defect because of high leading angle and trailing angle.

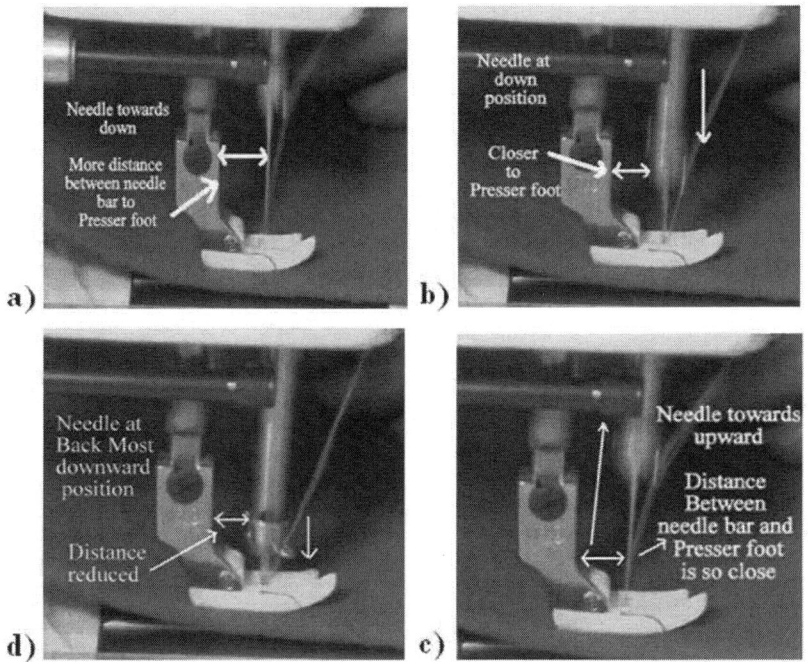

Figure 3.13 Working of needle feed mechanism.

- The main requirement of the needle feed mechanism is its usefulness in bulky sewing situations, such as when quilting through fabric and wadding. When three or more layers are under sewing process, the previously discussed fed mechanisms will be the ones which will be able to control the top and bottom plies.

- In the case of some stitching situations like laces on different fabric layers, there is wide possibility of slipping the intermediate ply from the feed dog and top presser foot holds. This situation seeks a positive control on the middle plies also.

In this mechanism the needle enters the material and remains in it while moving perpendicular to the needles (normal direction of travel), thereby feeding the goods or assisting in feeding the goods. This assists in preventing upper, middle and lower layers of material from slipping in relation to each other. It is often used in conjunction with drop feed and/or with upper feed.

There are three types of needle feeding:

- Upper pivot needle feed
- Central pivot needle feed
- Parallel pivot needle feed

3.2.5.1 Upper pivot needle feed

- The needle bar, which carries the needle, is held in a frame, and its motion is pivoted from a point on the frame farthest from (or far from) the needle.

- The needle enters the goods at a certain angle and exits at a different angle. This would seem to disrupt the material and the stitching process, but in practice it does not. (The needle will enter the goods at a leading angle from the centerline of needle travel, and will exit the goods at a trailing angle from the centerline of needle travel).

3.2.5.2 Central pivot needle feed

- The needle bar, which carries the needle, is held in a frame, and its motion is pivoted from a point near the middle of the frame.

- The needle will enter the goods at a greater leading angle from the centerline of needle travel than with the upper pivot system described above, and will exit the goods at an equally great trailing angle from the centerline of needle travel that it entered with.

- There is lesser momentum of the needle bar frame in motion than with the upper pivot system and higher stitching speeds can be reached.

3.2.5.3 Parallel drive needle feed

- The needle bar, which carries the needle, is held in a frame and its motion is always parallel in relation to its prior and successive movements. It remains perpendicular to the material at all times.

- For example, if the needle enters the goods at 90° to the materials' surface, the needle will remain at 90° through its travel and will exit at the same 90° angle.

- It is the type of needle feed suitable for stitching the heaviest and thickest of materials. The mechanism involved in a parallel drive makes for a more expensive unit and will generally have a slower stitching speed.

3.2.6 Compound feed

Compound feed mechanism is a combination of synchronized drop feed and needle feed mechanism, as shown in Figure 3.14. The material is fed through the combined motions of the needle and feed dog. Compound feed utilizes a feeder below the plate that rises up through the plate, compresses the fabric against the presser foot in conjunction with a feeder above the plate and needle would be found penetrating inside the fabric. At this juncture both needle and feed dog will move one stitch and the fabric will move along with them.

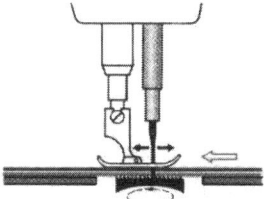

Figure 3.14   Compound feed mechanism.

The needle will be in the fabric while feeding is taking place and the plies of the fabric will be held together. It is particularly useful in bulky sewing situations, such as when quilting through fabric and wadding. Exact synchronization of the movement of the needle and the drop feed is needed. The needle hole is generally within the feed dog. Stitch length adjustment is normal with the needle stroke changing with the stroke of the feed dog.

3.2.7 Unison feed

Unision feed mechanism is a combination of needle feed and top feed mechanism. In this mechanism, as mentioned in top feed mechanism, the presser foot used will be of **two-part**, the centre part of the feet will move with the needle as mentioned in needle feed which is shown in Figure 3.15.

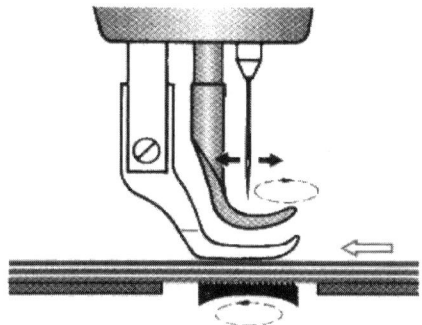

Figure 3.15   Unison feed mechanism.

The mechanism holds positive control not only to the top and bottom plies but also for intermediate plies. However, the applications are limited because once needle feed is included in a system which also has variable top and bottom feeding; it removes any opportunities for adjusting the amount of feeding in the top and bottom plies by means of the separate upper and under feeds, or by the operator easing in. It is valuable in the sewing of certain problem materials, especially those with slippery or tacky surfaces. It can achieve seam joining and edge binding without ply shift.

In Figure 3.16(a) it can be seen that the needle and top feeder enter into the sewing action from the top most position of the needle, while the main foot will be in action by holding the fabric. Figure 3.16(b) shows the needle's down most position, till this stage, it can be seen that there is a considerable distance between needle bar and the presser foot bar.

In Figure 3.16(c), the needle and feeder start moving to the backward most position, once the sewing cycle is over along with the fabric and feed dog. This can be noticed in the figure with the zero distance between needle and presser foot bar. At this stage the main foot disengages the fabric and raises above. In Figure 3.16(d), the needle bar along with top feeder moves up, back to the starting point, where the main foot will be back in action to hold the fabric.

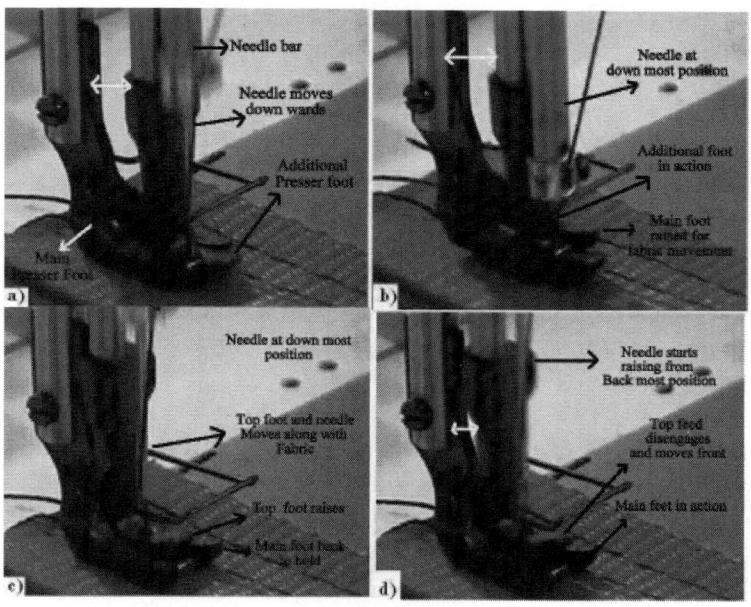

Figure 3.16   Working of unison feed mechanism.

3.2.8 Puller feed

- A puller feed is a way of providing positive control of all the plies of fabric as they leave another feeding mechanism such as drop feed as shown in Figure 3.17.

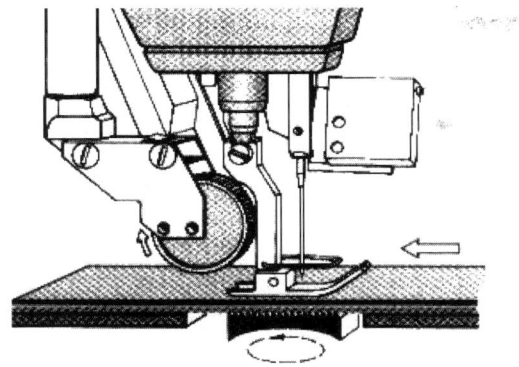

Figure 3.17 Puller feed mechanism.

- One or two rollers exert a pulling motion on the fabric immediately behind the presser foot. Both rollers or the top roller only may be driven while the lower one is idle as shown in Figure 3.18.

Figure 3.18 Puller feed mechanism working.

- Puller feed is particularly useful in multi-needle stitching of parts such as waistbands and it may be set slightly faster than the main machine drop feed to overcome any tendency for the seam to twist.

- It has been developed for the insertion of sleeves on tailored jackets and it can be programmed to insert different amounts of ease in the different sections of the sleeve head by adjusting the speed of the top belt compared with the bottom belt.

- A secondary top belt to the right of the main one can move faster on curves and assist in taking in the excess material in the seam allowance compared with that along the stitch line. The belts have a lifespan of several years, provided they are not allowed to run against each other with no fabric between them.

3.2.9 Wheel feed

Wheel feed utilizes a roller that advances the fabric one stitch length at a time in a ratcheting motion. The presser foot has small rollers to permit easy movement. Wheel feed is used when there is a chance for the material being sewn to be damaged by tooth feeders. Examples are vinyl plastic and some leather products.

- A rotary wheel with a movement in the direction of feed.

- Incorporates a friction surface or clamping surface that feeds or assists in feeding the goods as shown in Figure 3.19.

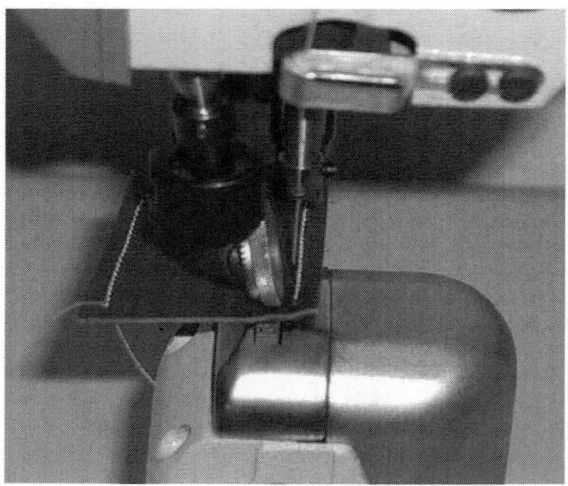

Figure 3.19   Roller / wheel feed.

- Has either an intermittent motion or a continuous motion.

- Continuous motion wheel feed must work in unison with a needle feed.

- Applicable to all types of sports shoes, shoes, children shoes, boots and other footwear and handbags sewing.

3.2.9.1 Upper and lower wheel feeds

It is the wheel feed system where both upper and lower wheels are driven. The material is fed between the wheels. There is positive feeding as pressure is applied on both the top and bottom of the material at the same time. The machine is shown in Figure 3.20.

Figure 3.20 Upper and lower roller feed.

3.2.9.2 Rotary wheel feed

It is a mechanical foot with teeth or friction surface or a smooth surface that assists in the transport of material. It can be considered as a kind of presser foot as shown in Figure 3.21(a) or separate feeder as in Figure 3.21(b). Often a foot working in unison with a drop feed has a directional movement in which the material is fed. This extra element may push downwards into the material to capture material between itself and another feed component or against the bed or a plate on the bed.

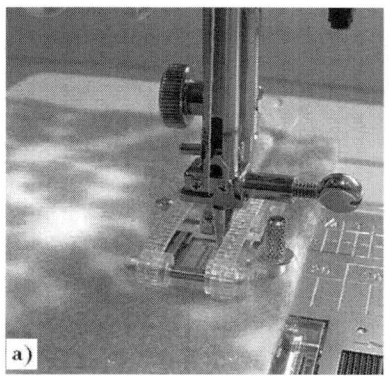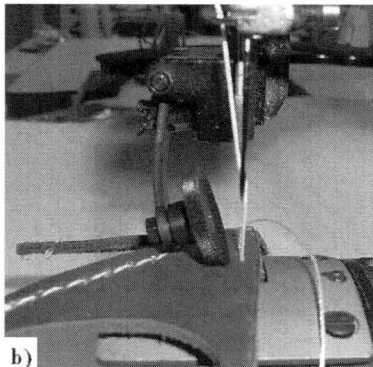

Figure 3.21 Rotary feeder mechanism.

3.2.10 Cup feed

Cup feed utilizes one or two cup shaped wheels that hold the edge of the material, permitting the needle to sew across the edge of the material. Often called a 'fur machine', as this machine is ideal for sewing the narrow strips together to create a fur coat. The machine is shown in Figure 3.22.

Figure 3.22   Fur machine or cup feed mechanism.

3.2.11 Clamp feed

A clamp feed clamps the material from above and presses downward, effectively holding the material between itself and the machine bed or a clamp, as shown Figure 3.23.

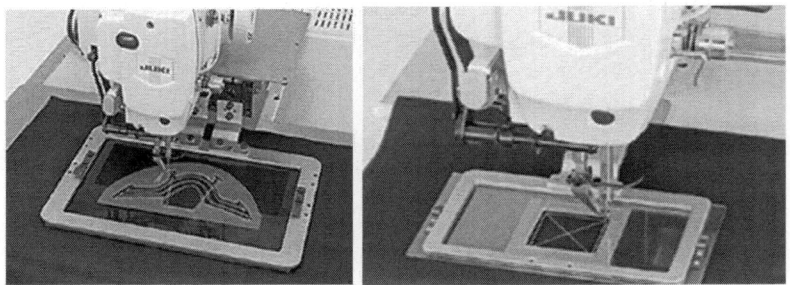

Figure 3.23   Clamp feed mechanism.

The clamp or clamp set is driven by linkage. The linkage moves the clamp and material under the needle as the stitches are being formed. The movement of the clamp and material can be in any direction or follow any pattern provided for in the linkage drive mechanism.

3.2.12 Machine speed, rate of feed and stitch size regulation

The machine speed and the stitch size decide the rate of feed. The relationship can be derived as follows:

$$\text{Rate of feed} = \frac{\text{Machine speed}}{\text{Stitch size (SPI)}}$$

Example:

Identify the rate of feed for the machine which runs at the speed of 2000 stitches per minute and forms 14 stitches per inch.

Rate of feed = 2000 SPM/14 SPI = 143 in./min

3.3 Sewing machine lubrication system

- The principle for lubricating sewing machines is the same as those for other machines.

- A lubricant is a gas, liquid or solid, placed between two mechanical links to reduce friction and heat that may be generated by link motion.

The important characteristics for sewing machine lubricants are oiliness, viscosity, flash point, volatility and pour test.

- Pour test – the lowest temperature of flow

- Oiliness – the degree of friction it possesses

- Flash point – the point at which the lubricant vapors ignite

- Volatility – rate at which the lubricant vaporizes with given heat

Out of commonly available lubrication oils like animal based, vegetable based, mineral and synthetic lubrication oils, the mineral oils are most widely used for sewing machine. Grease is the most popular solid used. Higher the speed of the machine, greater will be the friction and heat factor. Hence, high speed machine requires great amount of lubricant per unit time.

3.4 Types of lubricating system

1. Mechanical application

2. Wick application

3. Gravity trip

4. Bath

5. Automatic system (drip, splash, spray or stream)

3.4.1 Manual application

In this system, oil from oil can is poured directly to the link by the operator or company maintenance person in a regular interval basis. It is poured either hourly once or after every 4 h or after every shift based on the machine working condition.

3.4.2 Wick system

- In a wick system (Figure 3.24), the link receives oil from a contacting wick. The wick receives its oil from a reservoir filled by manual application. Wick and reservoir vary in size.

- The greater the lubricant need. greater should be the reservoir and wick size.

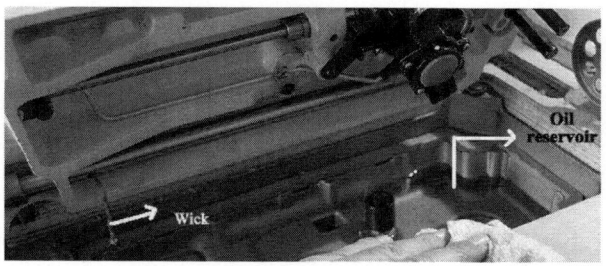

Figure 3.24   Wick systems for lubrication.

3.4.3 Gravity trip

- This system consists of reservoir with a valve device. The valve can be regulated to give desired drip rate.

- The higher the machine speed, greater will be the drip rate. The drip rate may be increased if changes in working condition require increased lubricant.

- Machine operating in a room with a temperature of 30°C requires more oil than a machine operating in a 20°C room.

- High room temperature slows the rate with which a machine may dispel the heat it generates. Most of these mechanisms are transparent and easily seen by operator. This helps the operator or maintenance person to refill the tank before it gets dry.

3.4.4 Bath system

- In this system, the whole sewing machine links and connections at the bottom will be immersed into an oil reservoir.

- It has marks and whenever the content reduces below critical level, the sewing machine oil would be replaced by the maintenance people.

3.4.5 Automatic system

Automatic lubrication system can follow any one of the mechanism which is described earlier except manual lubrication. Drip, splash or spray methods can be used or the type can be decided based on the machine activity.

3.5 Summary

This chapter gives an overview of sewing machine feed mechanism and their applications. The different types of feeding mechanism and their working principle are explained in depth for the better understanding with detailed description. The second part of the chapter discusses the common lubrications systems used in sewing machines and their working.

References

1. Carr, H. and Latham, B., *The Technology of Clothing Manufacture*, 2nd edn., Blackwell Scientific, Oxford, 1994.

2. Cooklin, G., Hayes, S. and McLoughlin, J. *Introduction to Clothing Manufacture*, 2nd edn., Blackwell Publishing, Oxford, 2006.

3. Solinger, Jacob, *Apparel Manufacturing Handbook-Analysis, Principles and Practice*, Columbia Boblin Media Corp., 1988.

4. http://www.industrialsewmachine.com/webdoc3/feed.htm (Accessed at 20th October 2014).

5. http://www.tolindsewmach.com/walking-foot.html (Accessed at 20th October 2014).

Sewing machine attachments

In this chapter various sewing machine attachments are classified and detailed based on their working and functions. The main attachments like guides, positioning attachments and finishing attachments are illustrated with examples for better understanding. The last part of the chapter describes various types of presser foots and their applications as an attachment.

Key words: Attachments, Guides, Folders, Finishing attachments, Presser foots.

4.1 Sewing machine attachments

Sewing machine attachments are the mechanisms which are attached to sewing machine for ease of working and are produced by the sewing machine manufacturer.

- The function of the attachments is to improve the qualitative and quantitative outputs of a sewing machine.

- Attachments are classified according to the work cycle function.

- The mechanism which is not a part of the machine as originally constructed is considered to be a sewing machine attachment. Attachments in the pure sense of the word are devices that can be attached to the sewing machine without cutting through or changing the original frame of the machine.

- The removal of such attachment leaves the machine in its original condition.

Sewing machine attachments either guides, positions/prepares the fabric for the sewing operation at the machine/it prepares the finished operation for some future process. Attachments have different names which usually indicate the function they perform like hemmers, binders, piping attachments, shirring and gathering attachments.

4.2   Attachment classification

Attachments are classified based on their movements during the work and also based on their functions.

4.2.1 Attachment types based on their movement during sewing operation

Static

*   Static attachments do not move during the sewing work cycle. These are the stationery attachments fixed with the machine bed during the working of sewing machine. These attachments help the operator to perform the operation more accurately with less fatigue. The static guide attachment is shown in Figure 4.1.

Figure 4.1   Static attachments.

Dynamic

*   Dynamic attachments are the attachments which are moved by the operator during the work cycle. They are attachments like folders and guides for specific application. They can be engaged during the requirement and can be kept idle when not in use, as shown in Figure 4.2(a).

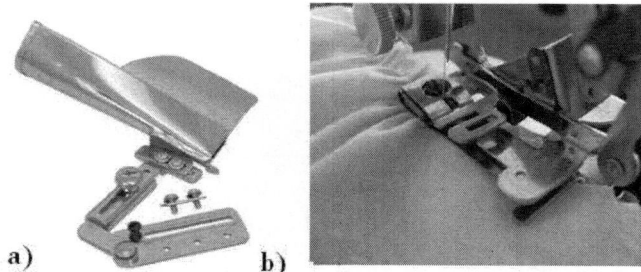

Figure 4.2   (a) Dynamic and (b) Synchronised attachments.

Synchronized

- These attachments have a link (or links) attached to the machine linkage system in order to synchronize it properly during the sewing action.

- This link is directly or indirectly connected to one of the drive shafts of the sewing machine.

- This moving link has a motion pattern which positions the fabric during the sewing element of operation work cycle.

- This pattern must be synchronized to the stitching action, the attachment driver is actually connected to the stitch mechanism driver. For example, shirring and gathering attachments shown in Figure 4.2(b).

4.2.2 Attachment classification based on their function

Table 4.1   Classification of sewing machine attachments

I Guide attachments	II Positioning attachments	III Preparation and finishing attachments
(a) Seaming attachments	(a) Seaming attachments	(a) Cutting
1. Seam with Guides *SSa 1 Seams and FS seams*	1. Headed seams *(lapped and super posed)*	Threads *(Stitch chain, Stitch thread)*
2. Trimming guides *Brides, Laces, cords, etc.,*	2. Bound seam 3. Piped seam	2. Fabric
(b) Stitching attachments	(b) Stitching attachments	(b) Pressing
1. Edge finish other than Hems and Bindings	Gathering . Shirring Tucking Hems	1. Pre sew
2. Decorative stitches other than edge finish	(c) Automatic positioners	2. After sew

4.3   Guide attachments

Guide attachments are the attachments which do not move or fold the fabric during the work cycle. The guide attachments are usually a focusing device (Figure 4.3) which enable the operator to position the fabric correctly and quickly during the operation. Guide attachments are gauge link attached to one of the three sections of the machine:

1.   The face

2.   The machine bed

3.   The presser foot bar

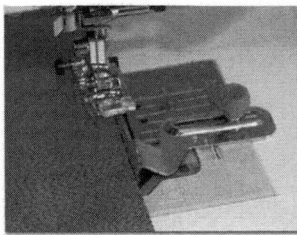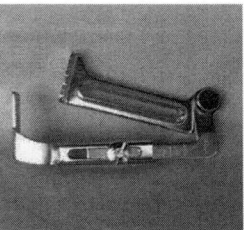

Figure 4.3   Guide attachments.

- These attachments present an edge or spot to the operator by which the operator uses as a focusing point. The operator does the positioning when the guides are used. Guides may be statics or dynamic. Dynamic guides must have a linkage system which permits the operator to temporarily remove the guide edge or point from the work area or line of vision during work cycle. This movement is done by either swivel or sliding action which is controlled by hand or leg action. Foot pedal and knee manipulators can be used to furnish the leg action. The length of the swivel or slide is a measure of the potential of the attachment.

- The ease of the swivel or sliding action is used to measure the efficiency of the attachment by the operator.

- The attachment guiding point may be shifted closer to, or farther away from the point where the needle contacts the cloth. There are 2 basic group of directions for shifting the guide edge closer to or away from the needle point:

1.   Horizontal: The distance is measured on the bed from the bed surface to throat plate needle hole. In this method, the guide shifts parallel to the machine bed, either to the side or front of the needle side.

2. Vertical: The distance is measured vertically from the throat plate needle hole. Here, the guide edge shifts at a vertical angle to the machine bed either to the side or front of the needle side.

Guides are also used to gauge seam and stitching width. Two placements may give the same qualitative factor (seam width) but one of the placements may be better for quantity of the production. Hence, the range of adjustment might be important quantitatively as well as qualitatively.

4.4 Positioning attachments

• Positioning attachments cause the fabric to bend, fold or shift during the sewing operation. Super imposed and lapped seam attachments are sometimes called positioning attachments. Positioning attachments are static, dynamic or semi-dynamic.

• They are dynamic because they fold the fabric during the sewing operation. For example, hemming attachments (also referred as 'felling' attachment). Attachments which position stripping (bias or straight), braid or ribbon around or on a fabric edge, seam edge a fabric surface or on a fabric surface or between fabric edges are all positioning attachments but these also called as **'Binding', 'Piping' or 'Stripping'** attachments.

• In binding attachment the stitching line, which sews the stripping, braid or ribbon around the fabric edge is seen on both sides of the binding. Bound seam types BSa, BSb and BSc are the examples of the three basic kinds of bound attachments, which are shown in Figure 4.4.

Figure 4.4   Basic bound seams commonly used in folders (a) BS a (b) BS b abd (c)BS c seams.

4.4.1 Right angle bias binder

Right angle bias binder is used for heavy and thick materials. These kinds of binders are used for the inside curving operations like neck-line, arm hole, etc. The attachments can be used for the binding of heavy weight (thick) materials from 5.5 mm to 6 mm. The outline of the binder is shown in Figure 4.5.

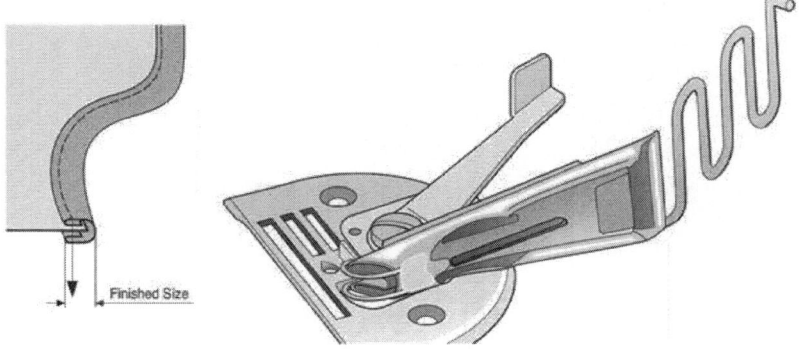

Figure 4.5   Right angle bias binder attachment.

4.4.2 Single-fold bias binder

Single-fold bias binder (Figure 4.6) is used for all kinds of materials. These kinds of binders are used for the outside curving operations and straight edges. The attachments can be used for the binding materials with a width ranging from 20 mm to 40 mm tapes and the finished width ranging from 5 mm to 12 mm.

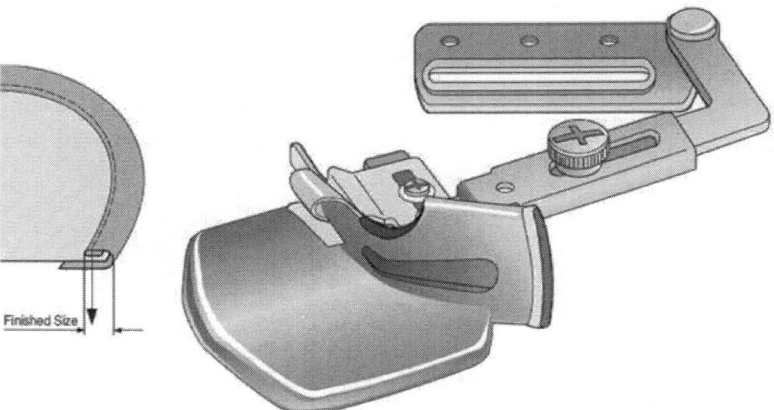

Figure 4.6   Single fold bias binder attachment.

4.4.3 Tape binder

This kind of binder is used for the straight edge finishing. The attachments can be used for the binding materials, width ranging from 10 mm to 25 mm tapes and the finished width range from 4.5 mm to 12 mm. The attachment is shown in Figure 4.7.

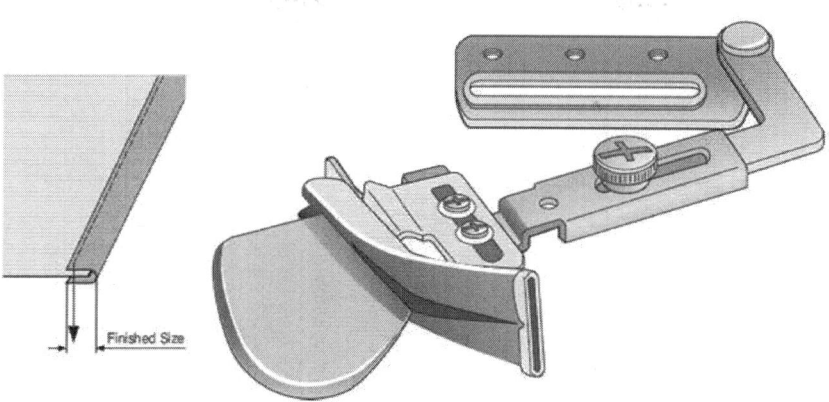

Figure 4.7   Tape Binder attachment.

4.4.4 Loop folder

The attachments can be used for the binding materials width from 20 mm to 40 mm tapes and the finished width range from 5 mm to 12 mm. The attachment is shown in Figure 4.8. This attachment helps in easy stitching on the corner of the other side of the belt by swinging the folder away and using the adjustable guide as a gauge.

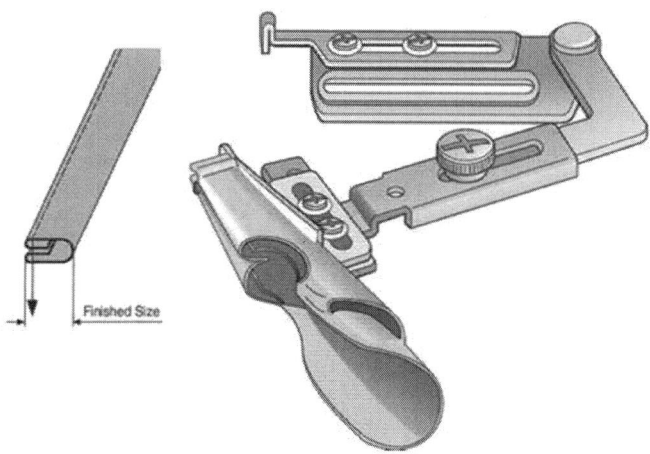

Figure 4.8   Loop folder attachment.

4.4.5 Straight folder

The straight folder attachments are shown in Figure 4.9, this attachment is used to attach the binding strip for straight line sewing. In order to increase the productivity, the base support has an extension. The width size ranges from 8 mm to 24 mm.

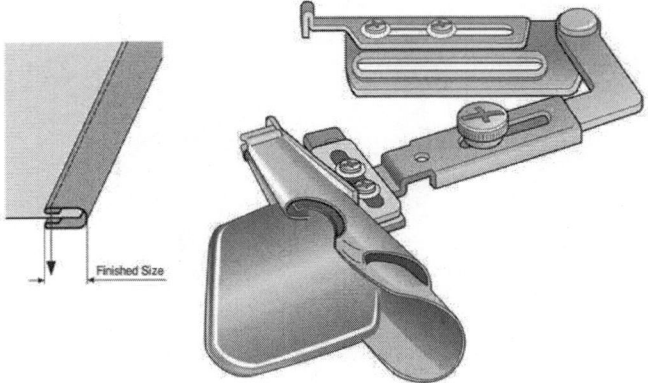

Figure 4.9   Straight folder attachment.

4.4.6 Belt loop folder

Figure 4.10 shows the belt loop attachment, commonly used in all kinds of sewing machines irrespective of the manufacturers. The finished width of the loop can be varied from 5 mm to 50 mm.

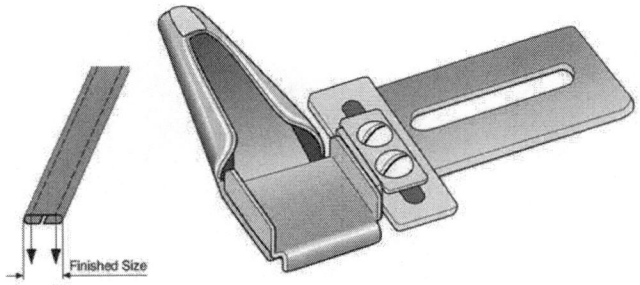

Figure 4.10   Belt loop folder attachment.

4.4.7 Cord edge piping attachment

Piping is an operation in which the folded stripping, braid or ribbon is inserted into a SS or LS type seam (Figure 4.12), with the folded edges of the tripping braid, or ribbon extending out of the seam as a decorative effect. Corded piping is a piping into which a cord has been inserted to round out the piping firmly, the attachment is shown in Figure 4.11.

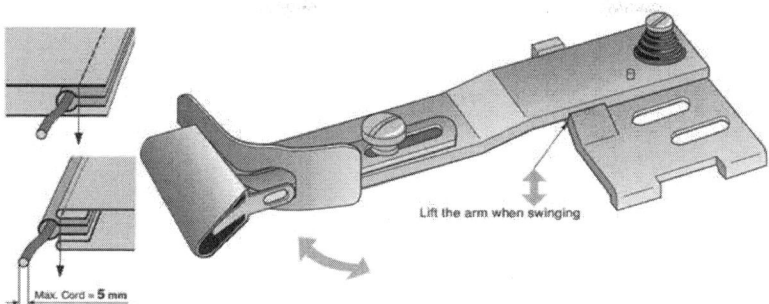

Figure 4.11 Cord edge piping attachment.

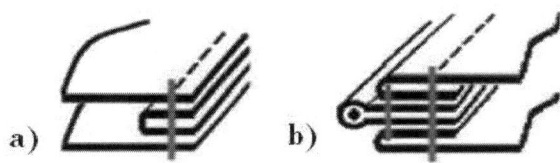

Figure 4.12 Seam types a) SS and b) LS Types.

4.4.8 Striping attachments

Striping is an operation in which the strip, braid or ribbon is sewed to the surface of the fabric, or as a cover over a seam or fabric edge, to be used as a decorative or reinforcement device or a finishing device in the case of some raw fabric edges. Seam types SSf, SSat, LSk and LSl are some examples for strip attachments, the seams are shown in Figure 4.13.

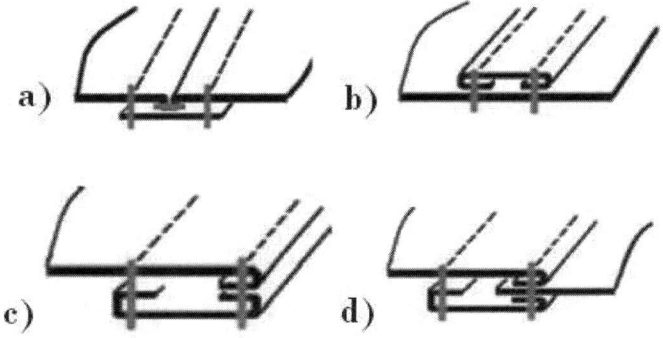

Figure 4.13 Seam types a) SSf, b) SSat, c) LSk and d) LS l.

4.4.9 Fell attachment – folders

Fell seam and felling attachments have curved links which fold the fabric. These links are called 'fellers'. A right-hand feller folds the fabric edge to the left as the fabric feed into the machine and vice versa. Most attachment manufacturers make three standard fell attachments with respect to fabric thickness.

- Light – for thinner fabric – attachment name (**L**)
- Medium – for medium weight fabric – attachment name (**M**)
- Heavy – for thicker fabric – attachment name (**H**)

Thicker fabric requires more space between the upper and lower planes of the feller. The depth of the fell seam or felling attachment for a given seam or hem width will vary with the fabric thickness. The thicker the cloth, the greater should be the distance between the opposite turns of the feller. Therefore, M and L attachments that produce same size of the seam width and seam heading on their respective fabric will have different heights and widths for the fellers.

The folder/felling is used

- To improve productivity
- To improve or maintain quality standards
- To reduce training time of operator
- To reduce fatigue of operator

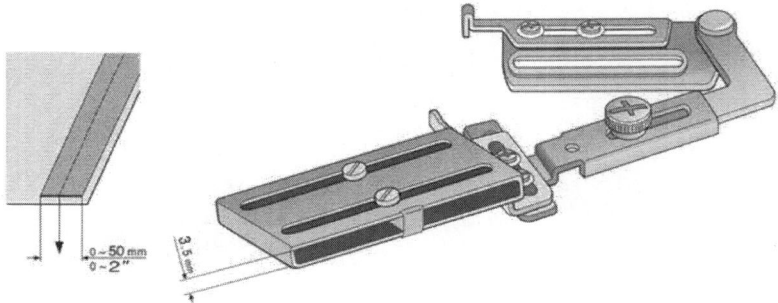

Figure 4.14   Adjustable striping attachment / tape guide attachment.

4.4.9.1 Swing hemmers raw edge – up turn

Figure 4.15 represents the attachment for raw edge up turn hemming operation. This attachment is also a dynamic one. The name swing hemmer implies that, the needle position (distance between needle point and edge) can be altered by adjusting the screw. The finished fell size can be varied from 5 mm to 25 mm (1 in.).

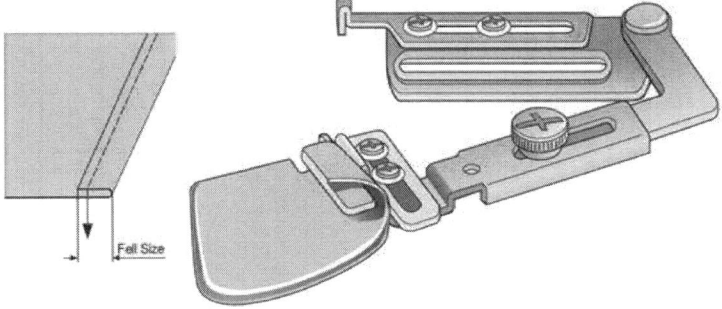

Figure 4.15   Hemmer attachment for raw edge upturn operation.

4.4.9.2 Single down turn feller

Figure 4.16 represents the attachment for raw edge down turn hemming operation. This attachment is also a dynamic one. The name swing hemmer implies that, the needle position (distance between needle point and edge) can be altered by adjusting the screw. The finished fell size can be varied from 5 mm to 25 mm (1 in.).

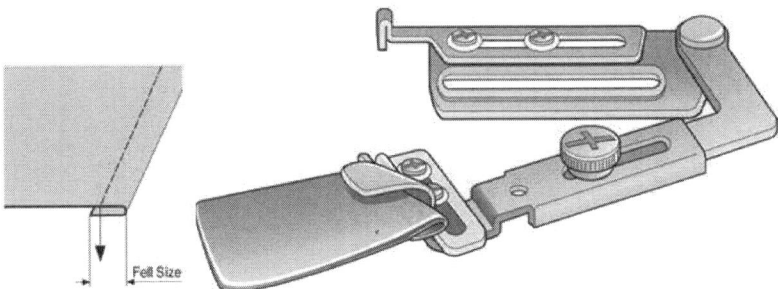

Figure 4.16   Hemmer attachment for down turn finish application.

4.4.9.3 Clean edge up turn hemmer

Figure 4.17 represents the attachment for clean edge up turn hemming operation. It is a dynamic attachment. It is known as clean edge hemmer because the raw edge of the hem is closed inside. The name swing hemmer implies that, the needle position (distance between needle point and edge) can be altered by adjusting the screws. The finished fell size can be varied from 6 mm to 35 mm (1 in.). The hemmer also available for extra width fell process from 40 mm to 50 mm, where, the different swing brackets will be used based on the requirement.

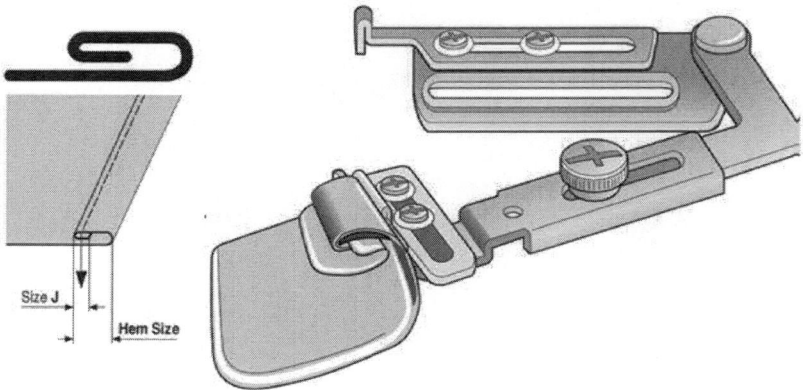

Figure 4.17   Clean edge finishing up turn hemmer attachment.

4.4.9.4 Latch elastic hemmer

Latch elastic hemmer attachments can be used for hemming operations with elastic tape inside. With this kind of hemmer, the width from 16 mm to 20 mm can be produced. The elastic width of 8 mm–10 mm can be used. The size J space of 7 mm–8 mm can be produced. Here the size J represents the space between the feller raw edge to fell inside fold as shown in Figure 4.18.

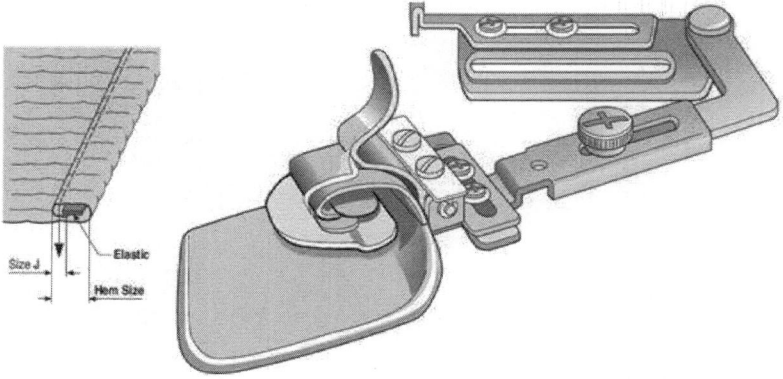

Figure 4.18   Hemming attachment for latch elastic application.

4.4.9.5 Clean finish long core hemmers

This hemmer is used for curtains, shirt fronts, sheets, pillow cases, etc. The long core provides full control at high speeds for continuous long run hemming. The hemmer is shown in Figure 4.19.

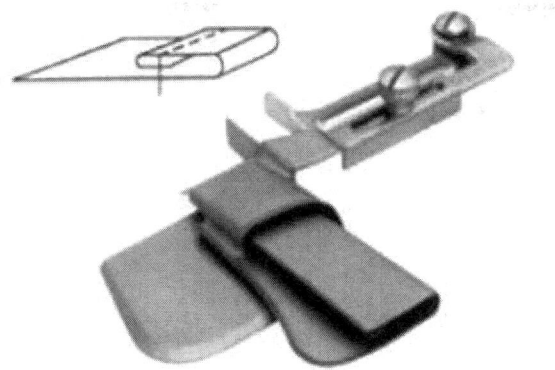

Figure 4.19   Clean finish long core hemmer for interior textiles.

4.5   Shirring and ruffling attachments

Shirring and gathering can be created with the presser foot, blade device and differential feeding mechanism. If the blade shirring mechanism is originally built in the sewing machine by the manufacturer, then that is not an attachment. It is still a positioner, but it is considered as a stitching auxiliary offered by machine manufacturer. Blade type shirring attachment shown in Figure 4.20 can be used for one of the following purposes.

1.   To gather the centre ply of three plies that are being seamed together but the outer plies are not to be gathered.

2.   To sew the gathers or shirring with a definite horizontal side pleat formation, such gathering or shirring is also called as ruffling.

Here, ruffling should not be confused with ruffles. Ruffle is a length of fabric that has one side free while the opposite side is gathered or shirred. Blade shirrers are moving attachments that are synchronized to the swing action of the machine by linking the driving mechanism of the drive shaft.

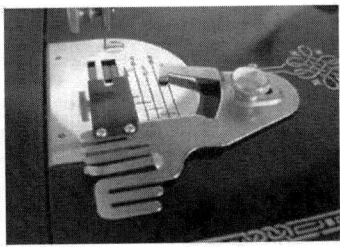 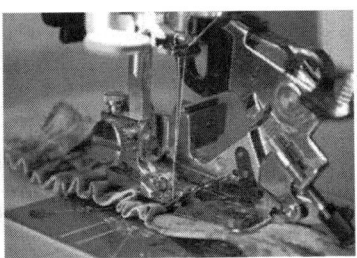

Figure 4.20   Blade type shirring attachment.

- Blade shirrer feeds the fabric that is gathered by pushing the fabric with serrated free end of the blade. The blade thrust must be longer and faster than the feed dog. Either of these two relationships causes the fabric that is pushed by the blade tip to be gathered or pleated in the stitching, this is shown in Figure 4.21.

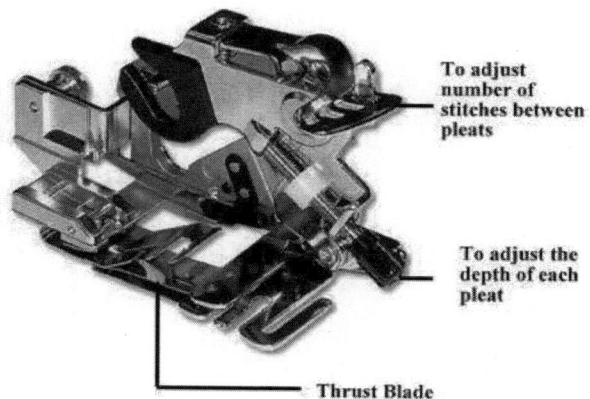

To adjust number of stitches between pleats

To adjust the depth of each pleat

Thrust Blade

Figure 4.21   Different parts of shirring attachments.

- The smoothness of the fabric surface decides the depth of the blade teeth. The smoother the fabric surface, finer will be the teeth, the contact angle will also vary with the fabric smoothness. Smoother the fabric, greater is the angle of inclination as it contacts the fabric.

- Fabric flexibility also controls the fabric reaction to the blade thrust. The more flexible the fabric, the smaller may be the blade tooth size and less angle of blade inclination.

- The ruffles can be altered based on the marking given in the ruffler. In Figure 4.22(b), the markings are indicated. The indications are follows.

If the selection lever is at,

- Star – regular straight stitch – no pleating

- Number 12 – a pleat every 12 stitches – least amount of ruffling

- Number 6 – a pleat every 6 stitches – medium amount of ruffling

- Number 1 – a pleat every 1 stitch – maximum amount of ruffling

Figure 4.22(c) represents the fabric passage flow though the ruffler. Here the broken line indicates passage of the fabric layer which needs to be ruffled.

The continuous line shows the bottom layer that does not require ruffle formation.

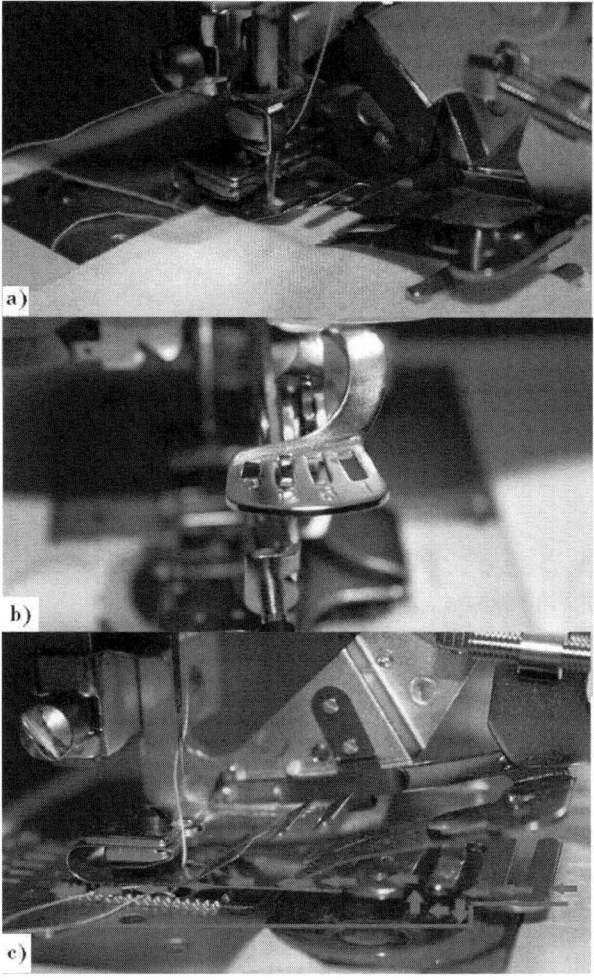

Figure 4.22   Shirring attachments (a) Fabric passes through shirring blade (b) Markings in shirring blade and (c) Passage of two layers through the shirring attachment.

Figure 4.23 shows the ruffles created at different markings and stitch lengths. Figure 4.23(a) shows the ruffles created at point 12 and stitch length 3 and 5. In the same way Figure 4.23(b) and (c) represents the ruffling foot setting point 6 and 1 with stitch density 3 and 5, respectively.

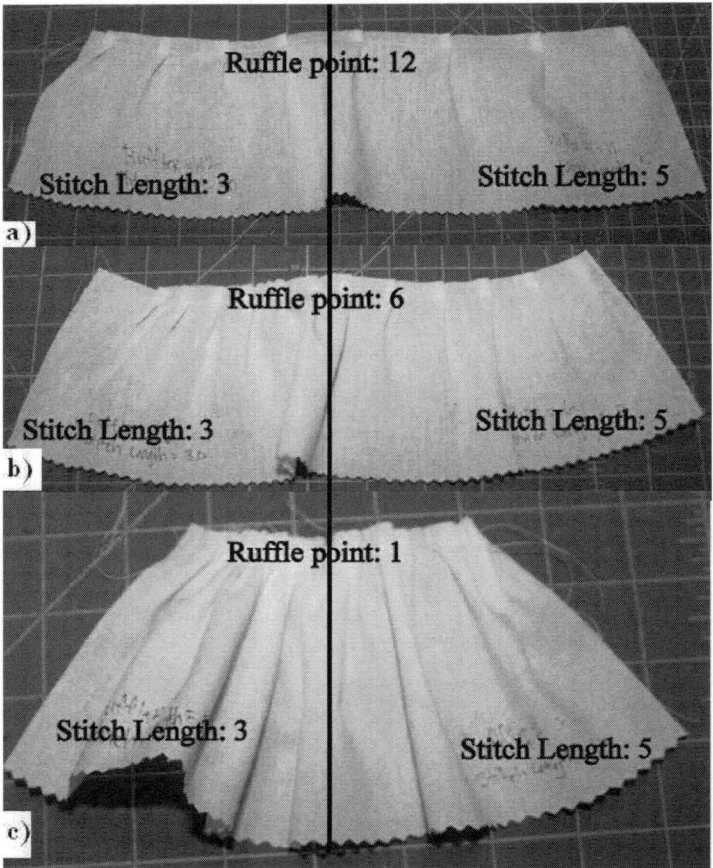

Figure 4.23   Ruffling created at different set points.

4.6 Tucking

A tuck is a fold of fabric stitched in place by running stitch or machine stitch on the right side of the garment. Tucks are used as a means of shaping the garment to the body, for holding in fullness or adding decorative effect at shoulders, waistlines, yokes, pockets or cuff of sleeves, etc. The tucks are of two types. They are explained in detail.

4.6.1 Knife tucks/pleat tucking

Knife tucking is also called as pleat tucking and is done with single needle lock stitch machine. The stitch types like OSf-1 and OSe-1 are the federal classification for the knife tucking (Figure 4.24).

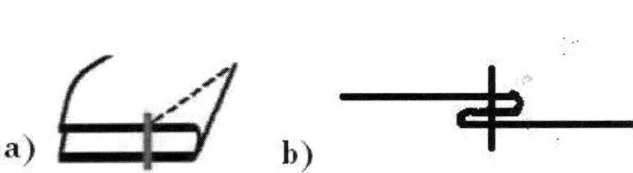

Figure 4.24   Tuck seam types a) OS f-1 and b) OS e-1.

The pleat/knife tucking attachments are shown in Figure 4.25 and the different parts of the attachments are explained in Figure 4.26.

Tuck guide – is adjustable and may be set for any desired width of tuck.

Tuck scale – contains Figures which indicate different widths of tucks. The tuck scale also acts as a smoother blade, keeping the tucks at a uniform width.

Tuck guide adjusting screw – the tuck guide may be set at any point on the tuck scale using this screw.

Space scale – contains Figure on the upper blade which indicates the width of the space between each tuck. The middle or 'grooved blade' contains a groove where the material is pressed by the spur at the end of the lower or 'spur blade', thereby marking the material for the folding of the next tuck.

Space scale adjusting screw – where the space scale may be set at any desired point.

Marking lever – presses on the groove blade, marking the material as it passes between the grooved and spur blades.

To fix the tucker in the sewing machine, raise the needle bar to the highest point, remove the presser foot from the machine and attach the tucker in place. The Figures on the tuck scale indicate the width of tuck in eighth of an inch; the marks between the Figures are sixteenth of an inch. The marks on the space scale are double the width of those on the tuck scale, so that both scales are set at the same Figure, blind tucks without spaces between them are made.

To make the space between the tucks, first the tuck scale must be set and then the space scale should be moved to the same number and as much farther to the left as required to have enough space. Each number on the space scale represents one-quarter of an inch and each mark between numbers are 1/8th of an inch.

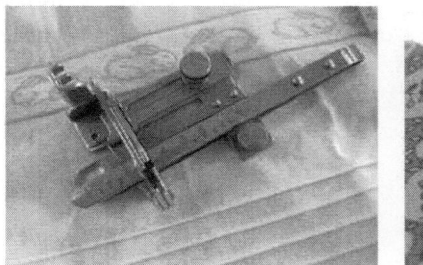 

Figure 4.25   Pleate / Knife tucking attachments.

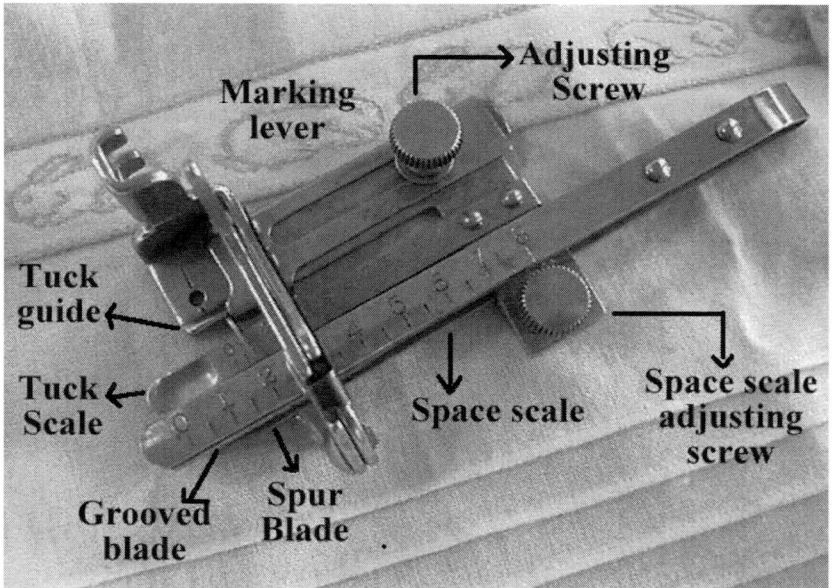

Figure 4.26   Parts of knif tuck.

4.6.2 Air tucks

Air tucking is done with multi needle machine. It is a ridge effect in the fabric plane.

When the cord is used as a filler in air tuck, the effect is called 'trapunto'. This effect can be obtained using specially designed presser feet. Stitch type Osc-1 is the federal classification for the air tucking and OSb-1, OSd-2 and OSd -3 are the federal classification for trapunto (Figure 4.27).

Figure 4.27   Federal classification of air tuck seams (a) OS b -1 (b) OS d-2
         and (c) OS d -3.

- Air tuck requires specially constructed presser foot and throat plate. The throat plate has a perpendicular ridge rising into the grooves of the sole of the presser foot. The height and width of the ridge and groove control the height and width of the air tuck.

- The average air tuck throat plate possesses a screw link system which permits one to rise or lower the protruding ridge. This allows one to adjust fine qualities of air tuck that are influenced by fabric qualities and characteristics.

4.7   Preparation and finishing attachments

4.7.1 Pinking

- Pinking is a common finishing operation. Usually pinking is done after SSa-1 seam is made.

- The pinking mechanism is attached to the machine as an attachment. Different pinking machines are shown in Figure 4.28.

Figure 4.28   Picking attachments for sewing machine.

There are two major pinking action used in pinking attachments

1. Chopping pinkers

   (A)   Oscillating vertical chop

   (B)   Vertical rotary chop action

2. Cutting pinkers

   (A)   Horizontal rotary pinker

   (B)   Vertical rotary pinker

4.7.2 Pressing attachments

Pressing attachments are used for finishing or preparation function after the material is sewed. Sometimes we do have a situation where the succeeding operation is harnessed to the end of the sewing operation.

For example:

- For a strap or belt operation which has a pressing device, either a flat iron or a rotary press attached to the machine head should be used.

- This attachment has a feed device, which feeds the belting or strip, in and out of the press.

- Many machines with belt, strap or belt loop attachment have fabric cutting. It cuts the sewed belting, strip and strap in to desired length. The length of the cut can be adjusted by changing the gear or cam size, it also alters the cutting knife cycle delay time.

4.7.3 Thread master

One of the finishing attachments, which come along with the cutting and pressing attachment, is the thread master. After the completion of sewing and pressing, it passes through the thread master attachment. It is an automatic electrical control device that stops the sewing machine when one of the following occurs during the sewing operation.

1. Thread breaks (needle, bobbin or looper)

2. Thread runs out from packages

3. Material runs out of feeding

4. Material tangles in between

These attachments can be used for gang machine operation in factories, where the machines can also be driven by line shaft in some cases.

These kinds of attachments are essential in the case of gang operation, where the single operator takes care of three or more machines at a time. The thread masters can be activated by hand control levers or push buttons. This helps the operator to attend the job easier and quicker.

4.7.4 Thread cutting attachment

Thread cutting attachments are essential in the case of industrial sewing machine. Since the speed of the machine is very high, the sewing thread waste percentage will be high if the operator hand tears the thread after the sewing operation. However, the hand tearing action is undesirable in any stitch chain because the operation may disturb the sewing quality by pulling the threads out of the end of the fabric. This causes the fabric to gather at the end of the sewing line. When these gathers are straightened, the end of the fabric remains not sewed, this causes seam open at the end.

In other cases like, stitch type 400 and 600 class machine, the stitch cannot be easily broken by the hand like 301 stitch, this further complicates the above mentioned problem. Or otherwise at the end of sewing, the operator needs to raise the needle and pull out the thread to trim it. Normally, therefore, when sewing with a lockstitch machine, the operator has to operate the pulley and/or to trim the thread with a pair of scissors when sewing a corner part of the material and at the end of sewing. When sewing with a lockstitch machine with a thread trimmer, the operator is not required to carry out the aforementioned actions since the machine automatically performs them.

There are two types of threads given below.

1. Automatic thread cutting attachment
2. Hand or foot operated thread cutter

Automatic thread cutting attachment

- These are specially constructed presser foot with a hinged knife which cuts the thread chain as soon as the fabric is sewed. The knives are synchronized with the sewing mechanism.

- This knife cuts threads when the presser foot raises the needle in its top most point.

Hand or foot operated thread cutter

1. Operator controlled thread cutters are used where more than one or two free stitch lengths are required at the end of the sewed fabric.

2. The operator cuts the thread at will. The length can be changed as required.

Working of thread cutter

While the lockstitch machine with a thread trimmer makes a half revolution from the 'needle-down stop position' to the 'needle-up stop position', the thread trimmer works to activate the trimming knife. The lockstitch machine with a thread trimmer, therefore, has (1) a thread trimming knife function for trimming the thread, (2) a needle position control function (stop with the needle up/down, etc.) and (3) a function for controlling the machine's number of revolutions.

Needle positioner places the needle in either upper or lower point of the needle thrust while the operator actuates the mechanism. The actuation is either by knee or foot pedal. This leaves the operator's hands free to manipulate the fabric quickly after the needle is positioned. Thread cutter mechanism is often used in conjunction with the needle positioner attachment when the needle is positioned at the top most position.

The thread trim device comes with the tacking attachment which automatically tacks the stitch at the start and the end to avoid the stitch chain from opening. There are two types of tacking, bar tack and back tack.

4.7.5 Automatic stackers

- Stackers are the attachments used for retrieving and stacking sewed parts in superposed fashion automatically, as soon as the part is sewed.

- This eliminates the need for the operator to extract the part after it is sewn.

- Thread trimmers are needed when stackers are used, to make sure that the threads making the stitch are held intact correctly to begin the next sew.

There are two basic kinds of stackers

1. Lift – in lift stacker, the sewed item is lifted by air flow into the stack.

2. Flip – flip stackers flip the sewed part over the stacking bar or platform.

In some flip stackers the bar is stationery and another bar flips the material as soon as it is sewed. In some cases the bar is dynamic, the bar movement flips the material after the completion of the sewing.

- Stackers can be actuated mechanically, electrically or by fluidic control.

- Stackers can be attached to any manually controlled sewing machine.

- Once it is attached, the stacking execution is automatic as soon as the operator finishes sewing a part. The sewing machine with stacker attachment is shown in Figure 4.29.

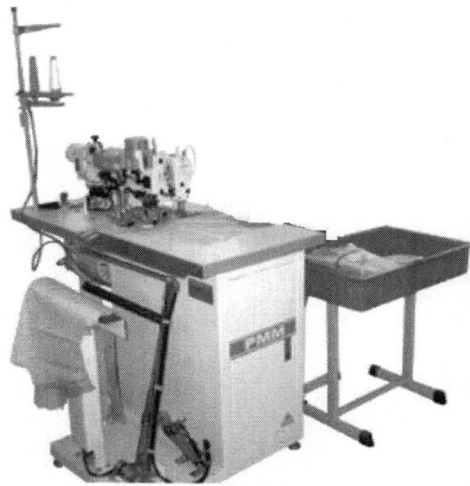

Figure 4.29   sewing machine with stacker attachments.

4.8   Presser feet

4.8.1 Functions of presser foot

- It acts as the upper part of the feeding combination that holds the fabric in place for the feeding action and stitch formation

- It is attached to presser bar and so it controls the amount of pressure placed on the fabric when it is fed through the machine.

- It assists in the operation of feed dog.

- The most common type is the flat presser foot. Figure 4.30 shows the parts of presser foot.

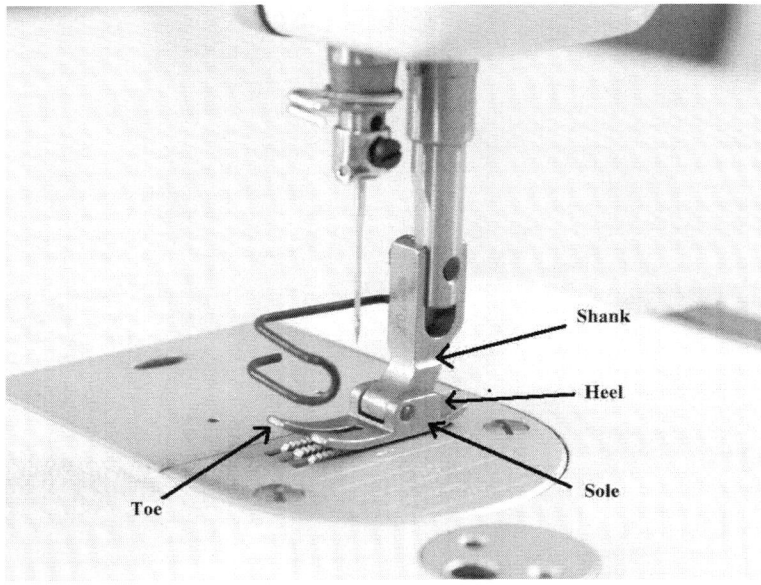

Figure 4.30   Different parts of Presser foot.

- Presser foot is a dynamic feed link. They are also kinetic links if they are part of a foot feed. The presser foot is attached to the pressure bar. The pressure bar in turn is connected by a spring action link which exerts pressure directly to the bar and indirectly to the presser foot.

- The spring compression is controlled by a thumb screw nut. This makes it easy to adjust presser foot quickly. Presser feet are either uni-linked or multi-linked. The section that contacts the fabric is the shoe. The contacting surface of the shoe is the shoe sole. The toe of the shoe is the area covering the fabric that is not sewed. The heel is the section which covers the sewed area. Soles usually are smooth surfaced, vary with respect to the shape and surface; which may be toothed, knurled or channeled. The section of toe or heel surface may be straight or curved. The front areas of many toes are curved. The toes and heels vary in length and width as well as shape.

- Presser foot comes in three basic types: long or high shank, short or low shank and the slanted shank. In short, the type of presser foot required by the sewing machine is dictated entirely by the needle bar. Depending upon the way a sewing machine is built, the needle bar can be long, short or slightly slanted. Then there are machines that are built to use snap on presser foot.

- A presser foot that is built for a machine with a long shank will not fit a machine with a slanted or short shank. If the sewing machine was designed to work with snap on presser foot, it will not work with foot designed for use on sewing machines with long, short or slanted shanks as shown in Figure 4.31.

Therefore, it is important to know the type of presser foot required based on the machine before investing in any additional feet.

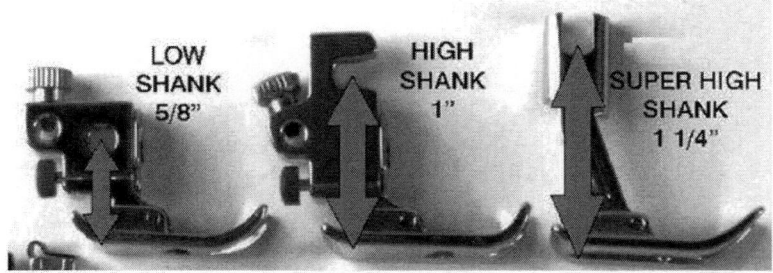

Figure 4.31   Classification of Presser foot based on shank height.

4.9 Types of presser feet

4.9.1 Compensating presser foot

- Compensating presser foot is used for creating an even stitch along the edge of the material. If the foot is left toe yielding type (Figure 4.32), then the left side foot will guide the garment to stitch on even edge to the left side of the sewing needle as shown in Figure 4.33.

- A presser foot may have more than one toes and one heels. A multi-linked presser foot has a hinge compensation or elevator compensation action or combined action of both. Hinge compensation is the action of tilting the sole plane by pivoting the shoe on a fulcrum, the hinge.

• Elevator compensation is the action of raising the entire sole plane. Both these actions are caused by the passage of unequal thickness under the presser foot during the work cycle. A presser foot may have toe, heel, or side compensation.

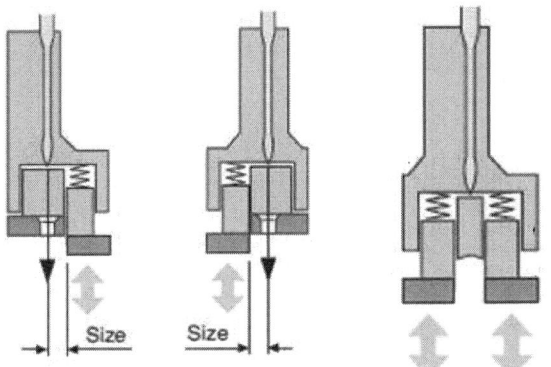

Figure 4.32   Compensating presser foot working mechanism.

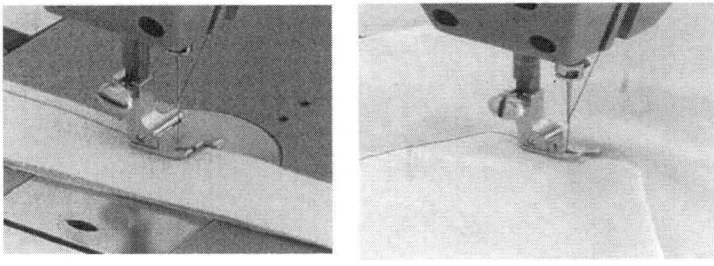

Figure 4.33   Compensating presser foot - Application.

Compensated presser feet are classified in to three types (Figure 4.34), they are

1. Left compensated foot

2. Right compensated foot

3. Double compensated foot

• The compensating presser foot size (as mentioned in Figure 4.32) mentions the distance between the compensating toe and the needle penetration point. There are different sizes available in the market ranging from 1.0 mm to 13.0 mm.

• The toe length variation is also available in this kind of presser foot, which is specifically used for the joining of small and complicated part. The toe length varies from 4 mm to 7 mm.

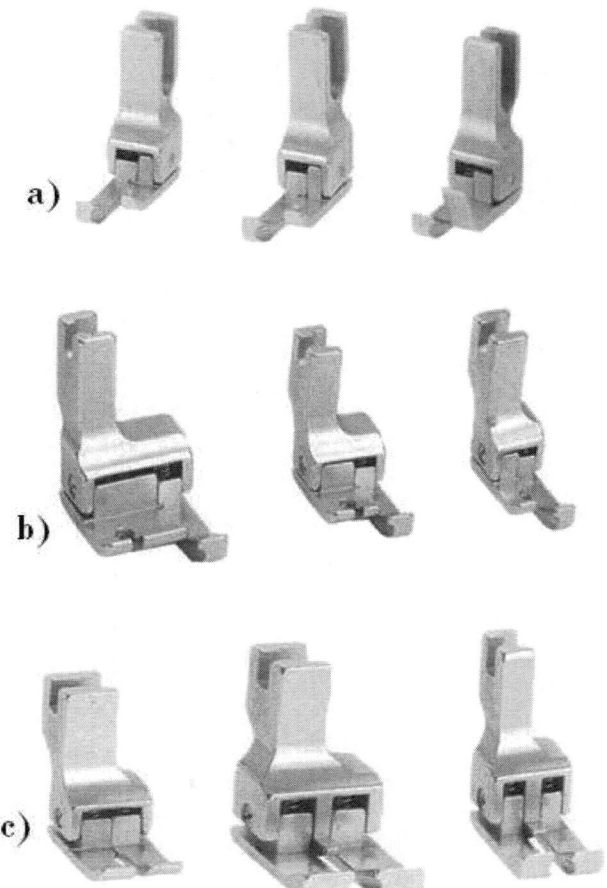

Figure 4.34 Compensating presser foot types (a) Left Compensated (b) Right Compensated and (c) double side compensated presser foot.

4.9.2 Teflon foot

This foot similar to the roller foot is designed to prevent fabrics such as leather and vinyl, plastic, suedes and ultrasuede from sticking to the bottom of the foot and to the foot plate of the sewing machine (Figure 4.35). This foot allows fabrics that are sticky to glide right under the foot. The foot unlike other foot has a distinctive look, like that of a white plastic material. This foot is also good for sewing zippers, allowing the foot to glide right over the zipper teeth.

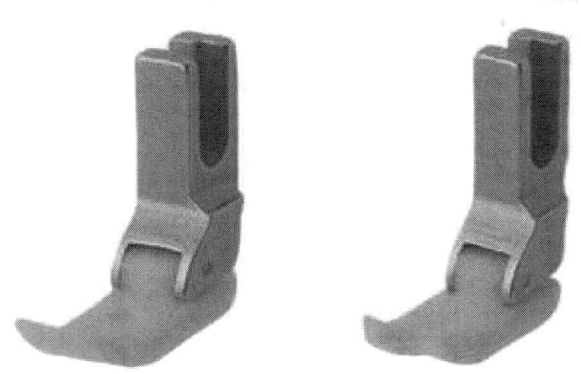

Figure 4.35 Teflon coated presser foot.

4.9.3 Gathering foot

The longer the stitch, the more gathered the fabric will be. The fabric that is underneath the gathering foot is the fabric that will be gathered. There is a slot to put another piece of fabric through. The fabric in the slot will not gather. This foot helps to form gathers/shirring effect in double layer stitching, where the bottom ply alone will gather and the top ply will be remain unaltered as shown in Figure 4.36.

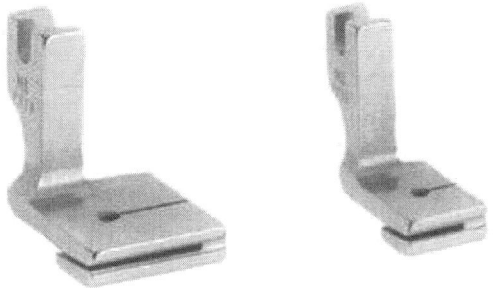

Figure 4.36 Gathering presser foot.

The gather creates soft gathers or a ruffle in the fabric as it is sewn, producing a slight gather and not a full-fledged ruffle. The setting of the stitch length controls the fullness of the gathers. When the stitch length is longer, the gathers are fuller; however, using a shorter stitch length decreases the size of the gathers. The type of fabric that is used determines the fullness of the gathers. For light weight materials such as silk, the gathers will be fuller and for heavy weight fabrics, the gathers will be lighter.

The gatherer should not be confused with the shirring foot, which is used to produce two or more decorative rows of gathers by parallel stitching, to decorate garments, such as sleeves or dress bodices; gathering by itself produces a single row of stitch. Both feet can be used to gather or shirr while attaching one fabric to another.

Figure 4.37(a) shows the gather formation in presser foot during the sewing of two layers. Figure 4.37(b) and (c) shows the gather formation at tension level 7, stitch length 3 and 5. Figure 4.37(d) shows the gathering effect created at tension level 5 and stitch length 5. Hence, the stitch length and sewing tension are the two major factors which affect the gather formation.

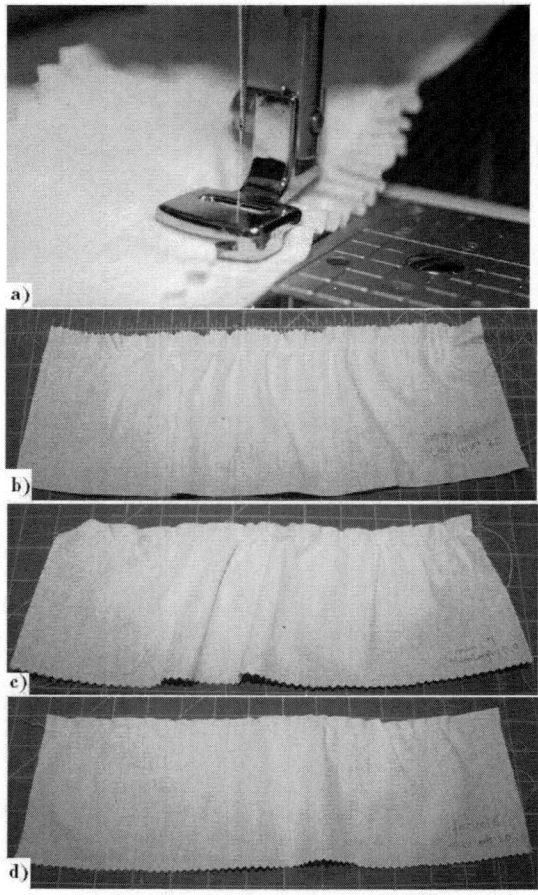

Figure 4.37   (a) Ply gather formation in presser foot (b), (c) and (d) Gather formation in different tension level and tension.

4.9.4 Piping/corded foot

Cording is the technique of binding or connecting an uncovered cord embellishment to a fabric. It is different from welting, which is covered cording, meaning that when the cord is encased inside of fabric, then it is attached to another fabric, usually for a heavy duty project. A pillow edge is an example for decorative effect.

Cording is also different from piping which is a lighter weight covered cord, placed inside of a seam or along a project edge as a trim on a garment. These are generally home decor techniques, but cording and piping can be used to decorate garments.

Sewing along besides the cording with the trough on the underside of the foot holding the cording in place is carried out. This foot comes with a variety of holes for insertion of cording used in decorative stitching and embellishing, which is why it is often called the multi-cord foot. These feet come with as few as 3 holes or as many as 9 holes. This is also a foot that nearly every sewing machine manufacturer offers. This foot can be used to create and attach cording or to attach pre-made cording, this is shown in Figure 4.38.

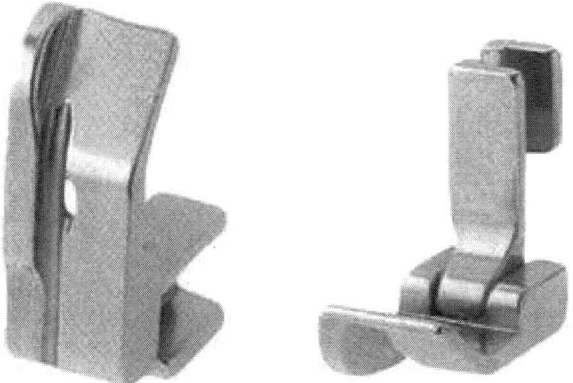

Figure 4.38   Piping or corded presser foot.

4.9.5 Edge guide presser foot

It is suited to apply lace and trims; edge stitch foot is also ideal for topstitching edges, pleats and hems. The guide blade in the centre of the foot helps to sew seams and hems running perfectly parallel to the edge very quickly as shown in Figure 4.39.

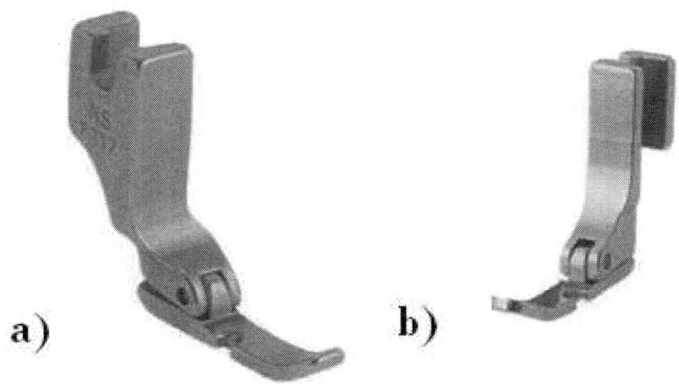

Figure 4.39   Edge gude presser foot.

4.9.6 Zipper presser foot

A zipper foot is able to adjust and so it is on the left or the right side of the sewing machine needle. This adjustment allows sewing the zipper without the presser foot applying pressure to the zipper teeth, which will not move when caught in the feed dogs because of pressure from a presser foot. A zipper foot is commonly used to sew trim that will not fit under the zipper foot. Using a zipper foot to sew on piping allows sewing next to the piping without the piping being jammed under the foot. Trims that would be damaged by the presser foot usually have a strip of fabric for attaching them. The zipper foot allows sewing on the strip without putting the presser foot on the trim. The zipper feet width usually varies from 2.5 mm to 3 mm as mentioned in Figure 4.40.

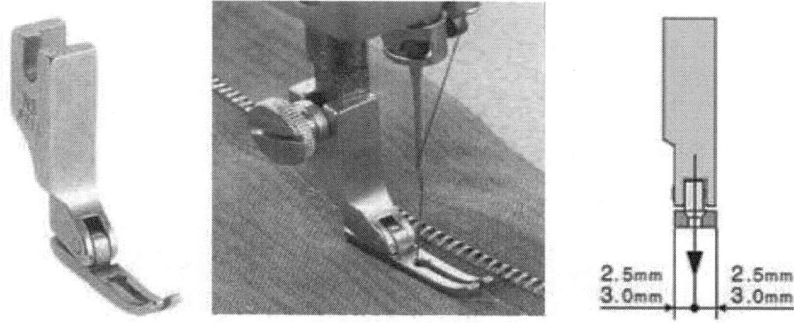

Figure 4.40   Zipper presser foot.

4.9.7 Hemmer foot

The width of the bottom groove determines hem width. Many narrow hemmers come in three widths – 2 mm, 4 mm and 6 mm, which denote the width of the groove on the bottom of the foot and the hem it produces. Different manufacturers call their hemmers 'narrow hemmers', others use the term 'rolled-hem' feet, and still others refer to them as 'shell hemmers', regardless of how manufacturers use these terms. The hemmer, regardless of manufacturer has the same structure basically; on the top of the foot there is a curved area that folds the fabric edge and a toe area to guide the outer edge of the hem. There is a groove on the bottom of the foot (Figure 4.41).

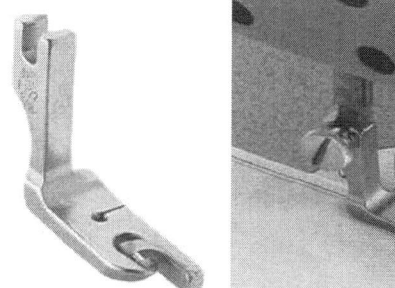 

Figure 4.41   Hemmer presser foot.

There are limitations in using this foot, the mechanism that makes the fold is limited and it is a rigid one, which limits the thickness of the fabric that can possibly be hemmed, making it possible to only hem light to medium weight woven fabrics.

A **narrow hemmer** is a foot of any width designed to fold a fabric edge over twice and stitch it down flat, with a straight stitch positioned along the inside fold.

A **rolled hemmer** also folds an edge over twice but it uses a centrally positioned zigzag stitch to pull the fold into a tight, round, rolled edge, similar to what vintage sewers refer to as a 'roll-and-whip' edge. This foot is typically 2 mm wide, with a rounded groove or channel underneath, instead of a flat channel as on narrow hemmers, as shown in anatomy of the narrow hemmer. Some rolled hemmers can also be used with a straight stitch for a tiny, flat hem.

A **shell hemmer** makes the same sort of fold as the others but it uses a zigzag or blind-hem stitch to form the outer fold into tiny scallops, called shells as shown in Figure 4.42. It can also be used with decorative stitches for other decorative hem effects. This foot may be available in various widths, especially if it can also be used for narrow hemming.

The given width of a foot indicates the size of the finished hem as shown in Figure 4.42, as well as the width of the opening in front and the channel underneath. Also, narrower feet cannot handle fabrics as heavy as wider feet can. Here is a listing of suitable weights for the most commonly available feet:

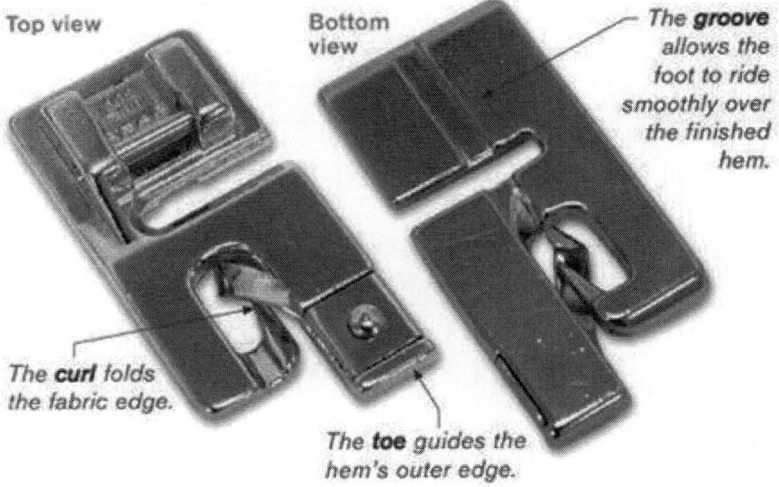

Figure 4.42   Parts and functions of presser foot.

- Rolled hemmer: For very lightweight fabrics only (handkerchief linen, batiste, voile, gauze). Makes hems approximately 1/16 in. to 1/8 in. wide.

- 2 mm narrow hemmer: For very lightweight fabrics. Makes hems approximately 1/16 in. to 1/8 in. wide.

- 3 mm–4 mm narrow hemmers: For light to medium-weight fabrics (broadcloth, shirting). Makes hems approximately 1/8 in. to 3/16 in. wide.

- 5 mm–6 mm narrow hemmers: For light- to medium-weight fabrics (poplin, flannel). Makes hems approximately 1/4 in. wide.

- Shell hemmer: For lightweight fabrics, including knits. Makes hems the same widths as narrow hemmers, up to 4 mm.

4.9.8 Quilting foot

The quilting foot is designed for free motion quilting and free-hand embroidery. While quilting, the drop feed allows you to sew curved stitches with ease. The vertical spring presses the foot securely to fabric being sewed, preventing the fabric from riding up with the needle (flagging). The material therefore lies flat while it is being sewn, allowing to produce neat, consistent stitches. For even greater sewing ease, quilting foot (Figure 4.43) features a clear sole, affording a perfect view of both needle and stitching area at all times.

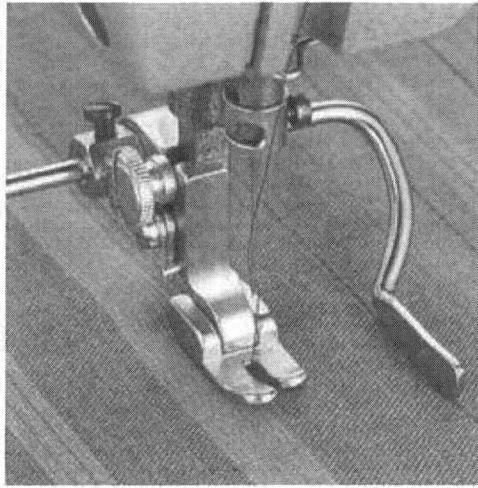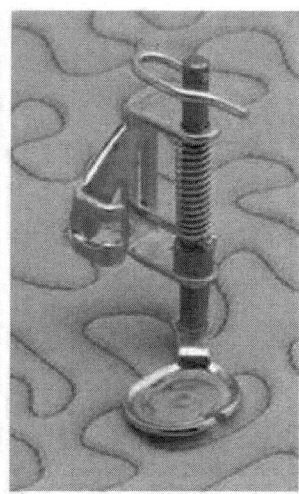

Figure 4.43   Quilting presser foot.

4.9.9 Pin-tuck foot

This foot has tiny grooves on the bottom and is used along with a twin needle. Fabric will tunnel under the centre groove and form a tiny fold (or tuck) in the fabric. The grooves on either side of the centre are used as guides for evenly spaced tucks that are meant to be close together.

Pin tuck foot helps in creating rows of delicate pin tucks for embellishing fine sewing. This foot is grooved on the underside of the foot with five, seven or as many as nine channels that are made for creating pin tucks. The foot is used with a twin needle also called a double needle. This foot is generally used for what is called heirloom sewing, given in Figure 4.44.

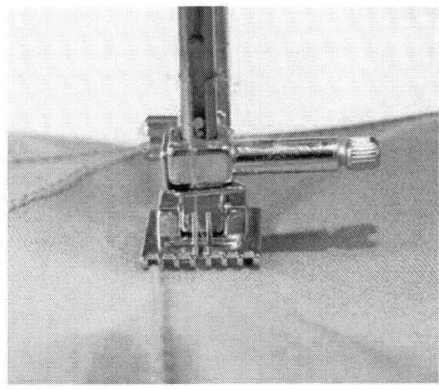

Figure 4.44   Pin Tuck presser foot.

The stitch is generally a straight stitch, with a stitch length of 2.0–3.0, and a twin needle size of 1.6 mm, 2.0 mm, or 2.5 mm. Pin tucks are generally sewn on light to medium weight fabrics, with the tuck in the fabric being created as it passes under one of the grooves. The rows of pin tucks are parallel, evenly spaced rows. The thread is a fine lightweight thread and unlike using a single needle, the twin needle uses two spools of thread.

3-Groove pin tuck foot (also known as the raised seam foot with raised seam plate)

• It helps in sewing quick tucks in medium weight fabrics.

• It is used with twin needle size 2.0 or 2.5.

• A straight stitch is selected with a length of 2.5.

5-Groove pin tuck foot

• It sews quick tucks in light to medium weight fabrics.

• It is used with twin needle size 2.0.

• A straight stitch with a length of 2.5 is selected.

7-Groove pin tuck foot

• It sews quick tucks in light weight fabrics.

• It is used with twin needle size 1.6 or 2.0.

• A straight stitch with a length of 2.0 is selected.

9-Groove pin tuck foot

• It sews quick tucks in very light weight fabrics.

• It is used with twin needle size 1.6.

• A straight stitch with a length of 1.5–2.0 is selected.

To create a more pronounced tuck, the top thread tension should be increased slightly so that the bobbin thread is allowed to 'pull' the fabric together more closely. This technique can be used when creating tucks without cording.

4.9.10 Binding foot

The bias binder foot is shown in Figure 4.45. This presser foot comes in two types, the standard bias binder and the adjustable bias binder. This is a useful foot for home decor sewing such as quilts and any other product that require a binding. The foot can be used with commercially made bias binding. There are also bias making machines on the market so that make creating new bias binding easy, as well as bias making rulers that assist the sewer in cutting the bias strips consistently with the same width.

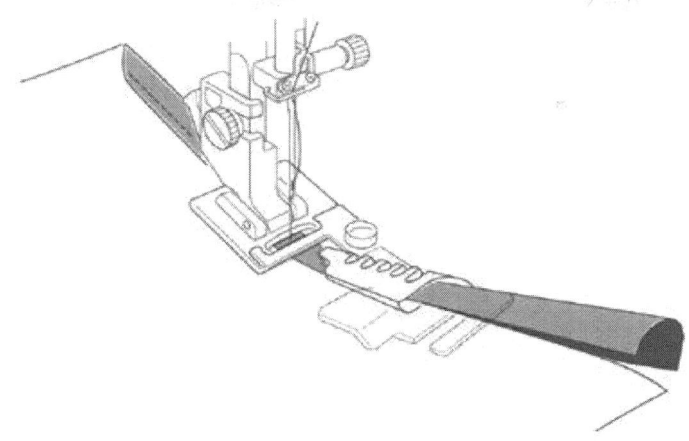

Figure 4.45 Binding presser foot.

4.9.10.1 Standard bias binder

The standard bias binder (Figure 4.46) is an interesting looking foot with a long cone protruding from the foot, which guides and folds in the edges on flat bias strips. The cone also holds single-fold bias tape in place as it is sewn. The width of bias strips that can be used for binding is limited to a maximum of 1 1/8 in.; this is the maximum width that can be fed through the foot's cone.

Figure 4.46 Standard Bias Binder presser foot.

4.9.10.2 Adjustable bias binder

The adjustable bias binder is a clear plastic foot with a screw and markings on the foot that allow for the adjustment of the width of the finished binding as given in Figure 4.47. This foot is generally used to fold double-fold bias tape to the edge of a project, such as a quilt. The width of the finished binding is determined and the foot is adjusted to achieve the desired width. The foot has two screws, one to hold the bias strip while it is being sewn; the second screw is used to adjust the foot to the right or left of the needle so that the stitching catches the edge of the binding.

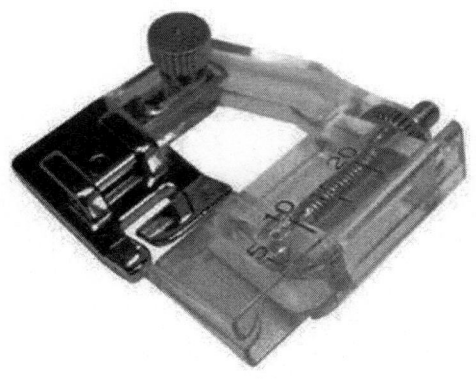

Figure 4.47   Adjustable Bias Binder presser foot.

4.9.11 Zigzag foot

To use a zigzag stitch, a presser foot with a wide hole and a needle plate with a similarly shaped hole should be installed. Without these, the needle will hit the foot and/or needle plate opening and break. For dense zigzagging or satin stitching, it should be made sure that the presser foot has an indentation on the underside to allow the stitch build-up to pass through easily. The zigzag stitch width and length can be adjusted to change the appearance of the stitch. The length refers to the distance between the stitches, and the width refers to the horizontal span. Depending on the machine, these may be pre-set options, or they may be adjustable, with larger numbers being wider and/or stitches farther apart. A typical zigzag foot is shown in Figure 4.48.

Figure 4.48   Zig zag presser foot.

- A zigzag stitch should lie flat against the fabric without tunneling. On some lightweight fabrics, it may be necessary to use a stabilizer underneath when sewing on a single layer. When used to stitch a seam, a narrow zigzag allows flexibility and it is particularly good for knit fabrics or wovens made with stretch fibres. Using a zigzag for seaming helps to eliminate thread breakage and subsequent tears in areas subjected to stress. Depending on the stitch width, it may be difficult to press seams open that have been stitched with a zigzag, so both seam allowances may simply be pressed to one side.

- To help control fraying, zigzagging can be used to finish seam allowance edges. This may be done on the single-layer seam allowance or on both together. Fabric may be trimmed close to a zigzag edge to prevent raveling.

A zigzag is often used for hemming stretchable garments, like knit T-shirts or pants. Because of its inherent flexibility, the hem is able to stretch without breaking stitches. When used on knits, a zigzag stitch can also create a lettuce hem. The hem of the garment should be stretched as it is being stitched. Combining a simple zigzag with decorative threads becomes an embellishment. These stitches are used singly or in multiple rows, it can create borders, stripes or other shapes for decorating garments or home decorating product. Zigzags also work well with double needles; machine instruction book should be checked for any width limitations.

4.9.12 Button hole presser foot

Buttonhole foot comes in a variety of styles, but it will help to create neat and uniform buttonholes on the garments. Those with a built-in memory allow to repeat the same size quickly and easily. This special foot does the entire buttonhole with no stopping and even knows how big to make the hole. In order to make the correct size buttonhole, the presser foot need to keep the button which is going to use into the foot. The plastic tab should be pulled as shown here until the button is snugly sitting in the foot (1): The button goes on the end and indicates the foot how big the buttonhole needs to be. A smaller button will move the placement of the plastic guides so the hole ends up smaller. The parts of the presser foot are shown in Figure 4.49.

To make a buttonhole, the button for which the hole is made is put in between the plastic holders. The quality of the foot is important, because there are buttonhole feet that are more manual and require positioning the needle from place to place in order to sew the next portion of the buttonhole.

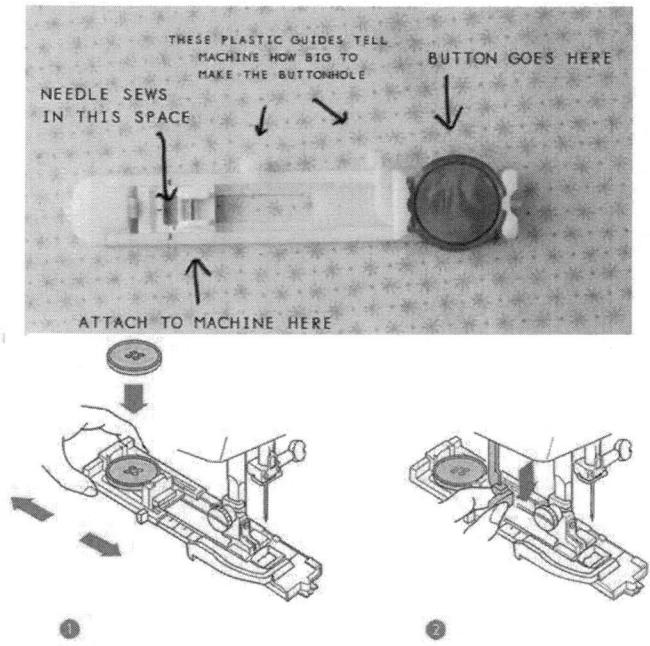

Figure 4.49   Button hole presser foot.

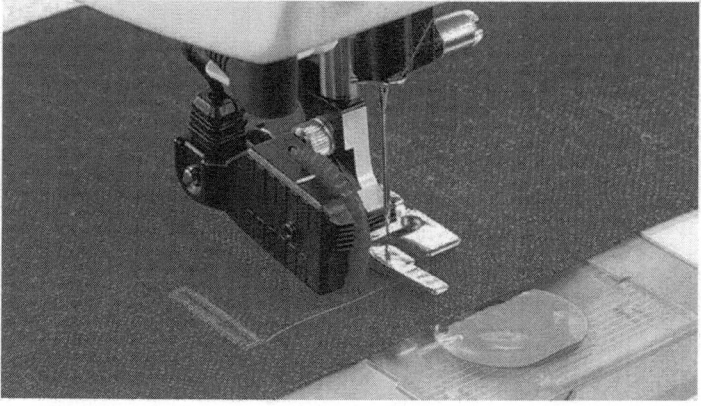

Figure 4.50   Button hole presser foot with sensor attachment.

One manufacturer has a sensor buttonhole foot that measures the buttonhole length and sews the size of the button that has been entered into the machine. There is no placing the button in a slot in the foot, the button is measured by the sewer and the measurement is entered into the sewing machine. This foot measures as it sews, creating every buttonhole with the same size. The newer foot makes creating buttonholes a lot easier. Although, the foot that came with the machine is the one that should be used, one can purchase a more automatic foot to make creating buttonholes easier (Figure 4.50).

The sensor one-step buttonhole foot measures buttonhole length and sew the button size that has been entered into the machine. The middle mark gives a 9/16 (15 mm) buttoning edge. The sensor one-step buttonhole foot measures as it sews, making every buttonhole the same size. The button size is adjusted in 2 mm steps.

4.9.13 Button attaching foot

The button sewing foot assists the fixing of the flat button. The foot actually attaches buttons and it can only attach flat buttons, it cannot attach buttons with shanks, but this is still a very useful foot. One version of this foot is a very distinctive blue and clear plastic foot, another version is an all metal foot, but they are basically shaped the same, how they are used to sew on buttons is basically the same.

The button sewing stitch is selected, and once the button is placed under the foot, then the needle is adjusted so that it enters each hole of the button as it moves from one hole of the button to the other by using the hand wheel to make sure that the needle enters each hole in the button cleanly before stitching begins. This is a key aspect so that the needle does not strike the button as it moves from side to side in stitching. For some sewing machines the feed dogs must also be lowered. Generally this foot is used with buttons that have two holes but it can be used with four hole buttons, the first two holes are stitched first then the button is repositioned so that the other two holes are stitched. The button attaching presser foot is shown in Figure 4.51.

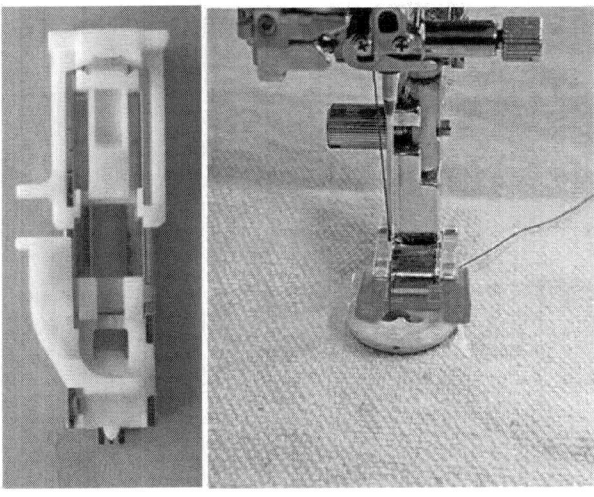

Figure 4.51 Button attaching presser foot.

4.9.14 Overlock foot

The overlock foot will neaten seam edges on knits and wovens on your sewing machine. The overlocking foot has many names; it is also called the overcasting and the over edge foot. The serger is a marvelous invention for finishing the edge of a sewing project to prevent the fabric from unraveling. This foot is used with the overlocking stitch that is part of the standard zigzag stitch that comes on sewing machines. The arrow points to the pin, which you align with the raw edge of your fabric. The pin keeps the fabric flat and allows the thread to wrap around it as shown in Figure 4.52.

Figure 4.52   Overlock / serging presser foot.

This stitch still cannot compete with using a serger but some sewing machine manufacturer have begun producing more complex attachments that can now give a serger a tough competition, by trimming the fabric as a serger would as it overlocks. The ability to overlock the edge of fabrics can be the difference to a professional looking versus a homemade looking garment or home decorative product; therefore, an overcasting foot is a great addition for sewing.

4.9.15 Edge stitch/blind hem stitch

The fabric should be folded as shown in Figure 4.53. The garment should be placed with wrong side up under the presser foot. The needle dropping position should be chalked and made sure that the needle is catching the fold of the fabric. The adjustable guide is used by loosening the screw on top of the foot, adjusting the plate right or left and the screw is tightened.

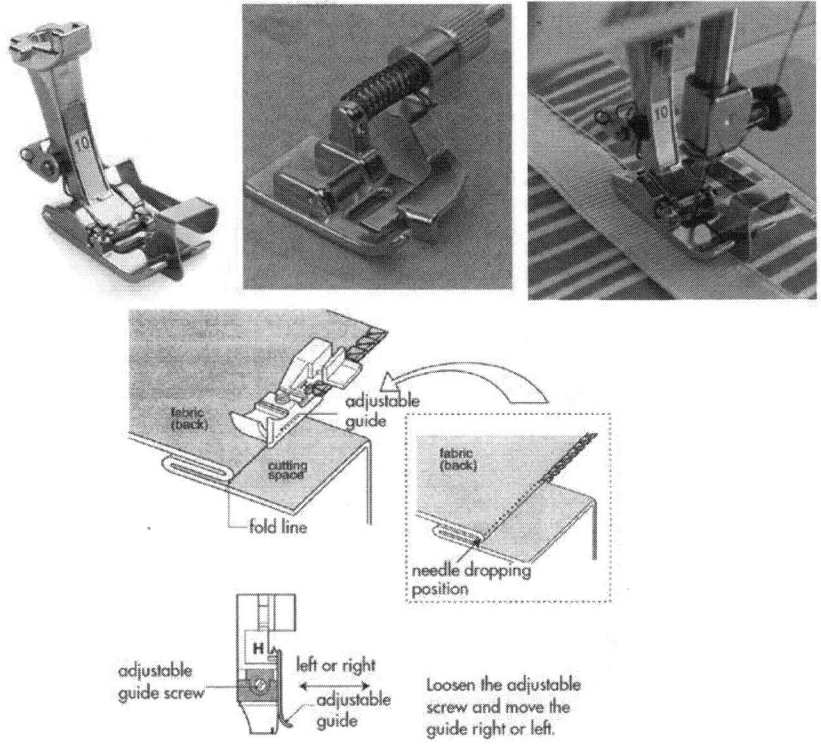

Figure 4.53   Edge stitch / Blind hem Stitch presser foot.

It allows aligning the blade along the edge of the fabric or on the seam line. The foot is designed so the sewer can top stitch or edge stitch a project. It has a 5 mm needle slot to allow for adjusting the needle position. These features are commonly found. The foot has a blade in the dead centre of the foot. This is exactly aligned with the centre needle position on the machine. This blade is a fabric guide which will be used to control the position of the fabric. It is similar to a 1/4 in. foot, with the same concept except it is in the centre of the sewing chassis bit. When the needle is taken a click or two to the left (or right), a perfect top stitching of 2 mm from the seam is achieved.

Tidy seam, stitch in the ditch

Sometimes the requirements want to press the seam allowance to one side and ensure they stay on that side of the fabric. To do this, seams should be pressed in one direction. Then edge stitch foot blade should be aligned exactly on the ditch of the seam. The needle should be moved to the side of the seam where fabric is pressed for seam allowance. Then it is stitched away and it will stay put and will give an even tidy finish on top.

Pleats and tucks

This foot can achieve lots of interesting texture with edge stitch foot. It can be used to repeatedly fold the fabric, stitch along the edge and then open up the seam. Doing so with the same needle position and even spaces results in tiny little pin tucks or pleat as shown in Figure 4.54(a).

Applique

The centre blade makes a wonderful guide while doing applique. It works with straight, zigzag and decorative stitches so it gives more versatility. A wide zigzag with fusible heat and bond can be used. The result centres the stitch across the edge of the applique object as shown in Figure 4.54(b).

Attaching fabrics together and lace

The foot can also be used to join things with folded or finished edges by using both sides of the guide and a very narrow zigzag stitch (Figure 4.54(c)).

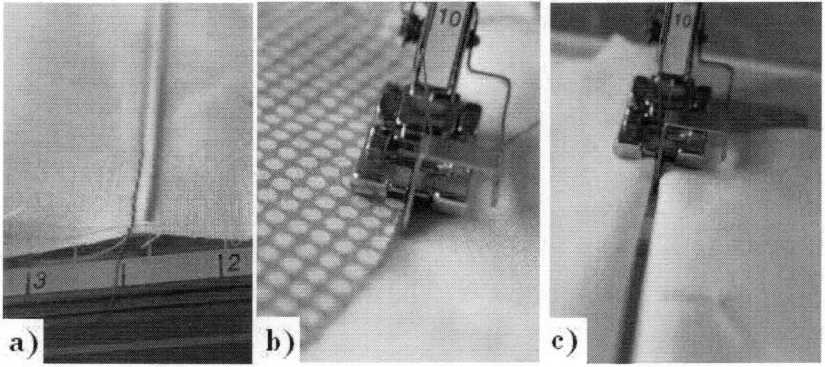

Figure 4.54   Applications of Edge stitch / Blind hem Stitch (a) Pleates and tucks (b) appliques and (c) fabrics or lace attachment.

4.9.16 Felling foot – lap seam foot

It is used for creating flat-felled seams or enclosed side seams commonly found on denim jeans, men's dress shirts and reversible garments. This foot saves time by doing the folding and pressing as the seam is stitched. The flat felled seam is the preferred seam in jeans, men's, women's shirts and children's casual garments. The advantage of a flat felled seam is that the seam allowances are encased within the seam.

The groove in the sole of the foot allows it to run and fell a 4 mm to 8 mm wide seam easily and precisely. The foot is well suited for medium to heavy weight fabrics such as denim, wool, corduroy and tapestry. For lightweight fabrics, 4 mm lap seam foot is recommended. The presser foot is shown in Figure 4.55.

Figure 4.55   Lap seam presser foot.

4.9.17 Invisible zipper

• The invisible zipper foot is not the same as the standard foot. This foot is designed to provide a genuinely concealed closure for garments and accessories. The foot looks different and works differently than the standard zipper foot.

• For one thing the invisible foot is used with the zipper open, and the right half of the zipper is placed under the right groove of the foot, the left half of the zipper is placed under the left groove of the foot. Two diagonal grooves act as a guide, making zip insertion very easy compared to the standard method. Located in the sole of the foot, the grooves easily accommodate the zipper coils, which are fed through at the same time as the zipper tape is guided under the foot and sewn as given in Figure 4.56.

 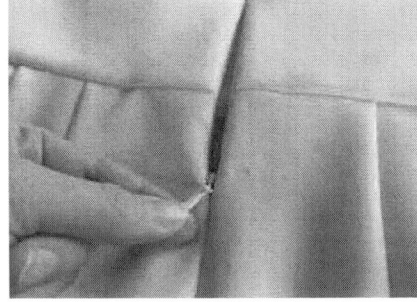

Figure 4.56   Invisible zipper presser foot.

- This is opposite of how the standard zipper foot works, which is designed to sew the right half of the zipper from the left side of the foot and turning the fabric to sew the other side of the zipper with the same side of the foot. Another difference is that the zipper is sewn to the garment before any other stitching is done to the seams.

- Another difference is that the zipper is actually sewn with right sides together, right side of zipper to the right side of the fabric and it is the fabric that is turned to conceal with zipper, once the stitching is done.

- The zipper is really considered invisible because there is no stitching showing on the right side of the fabric because the fabric has been folded to the inside of the garment.

4.9.18 Beading foot

- The beading presser foot is also called the sequin foot or pearl foot. The beading foot is used as its name implies, to sew beads or pearl strands onto garments or other projects. Generally invisible thread (also called monofilament thread) is used for the upper thread spool and thread to match the fabric is used in the bobbin.

- The stitch used is an heirloom applique stitch or overcast stitch is also called the zigzag stitch on some machines. The presser foot generally comes in two sizes, one with a small groove size for beads or pearls less than 2 mm and a foot with a larger groove for beads or pearls from 2.5 mm to 4.0 mm as shown in Figure 4.57.

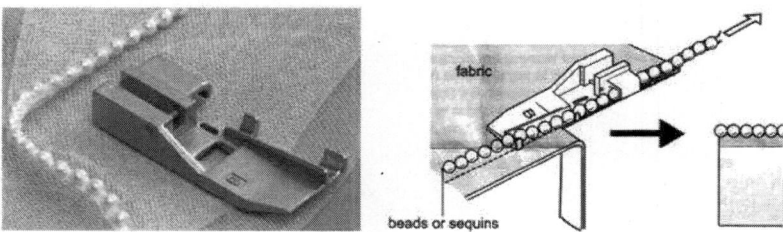

Figure 4.57   Presser foot for bead attachment.

- This presser foot is generally a clear foot with a groove under the foot in the centre, although some manufacturers offer a metal foot and generally those are the serger version of this foot. The clear presser foot makes it easy to see the beads as they are sewn to the fabric and it helps to see that the needle is properly clearing the beads. This groove is where the bead is inserted. Stabilizer is also placed under the fabric before sewing begins.

- Adding beads or pearl strands to a garment can add an elegant touch and is great for crafting and decorative embellishments. This foot is not one of the standard feet that comes with most sewing machines and is usually purchased as an accessory foot.

4.9.19 Darning foot

The darning foot (shown in Figure 4.58) is indeed a multi-purpose foot, not only it is used for embroidery but it is also used for free motion quilting and of course darning, which is also just basic, old fashioned garment mending. This foot can either be made of metal or plastic, but has a distinctive circular shape, regardless of whether the foot is made of metal or plastic. The foot can also be an open toe foot or a partially open toe foot.

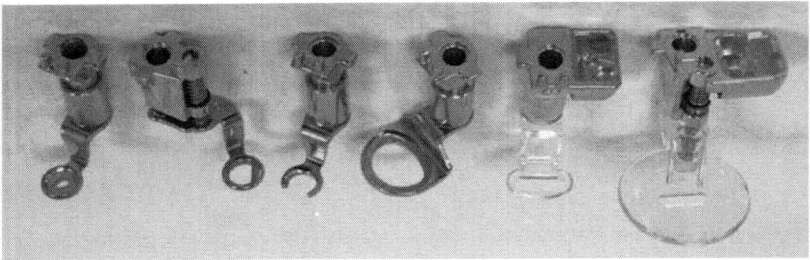

Figure 4.58 Darning presser foot.

The foot is actually attached to the presser foot bar by removing the existing presser foot and shank and attaching the darning/embroidery/quilting foot, which generally has its own shank. With this foot the feed dogs must also be either dropped or covered. If the foot does not have its own shank, then it is attached to the existing shank, like any other presser foot.

For embroidering and darning, the fabric should be hooped for stability. For free motion quilting, the fabric should be moved back and forth under the foot by the sewer as stitching occurs, the key to obtaining the desired stitches is moving the fabric at the correct speed, not too fast and not too slow.

4.9.20 Open toe foot – open toe applique foot/satin foot

Satin stitch best describes its most useful purpose which is to allow the buildup of stitches, to pass freely under the foot. The best way to find the difference is to look at the bottom and see if there is a wide groove or channel that runs the length of the foot. This channel allows the thickness of dense stitching to move freely under the foot. The satin stitch is most often used in applique (Figure 4.59(b)). The secret to good satin stitch is **tension** and **stitch length**. The new computerized machines automatically set the tension for satin stitch;

however in most cases it can be changed. This is especially important when using metallic threads.

To find the right combinations of tension and stitch length, enough time should be spent by the sewer for practicing on a scrap of the fabric that will be used with a piece of stabilizer backing it. The tension is correct when the **upper threads wrap around** to the back with a bit of it showing on either side or even one side of the bobbin thread. No bobbin thread should show on the top of the fabric.

This is an embroidery stitch where the stitch is worked close together in parallel across the fabric without any spacing, covering the entire area that it is used in. The stitch when completed is designed to look like satin, thus its name. This stitch is originally a hand embroidery stitch, which has now been adapted for the sewing machine. Satin stitches are generally found on just about all sewing machines as standard stitches. These stitches can be customized to fit a desired design by adjusting the length and width of the stitch to make these stitches look different.

The stitches created by this foot are the backbone of applique projects for machine applique and machine embroidery. The sewing machine uses the zigzag stitch to create these designs. In addition, for very light weight fabrics, stabilizer should be used on the back of the fabric.

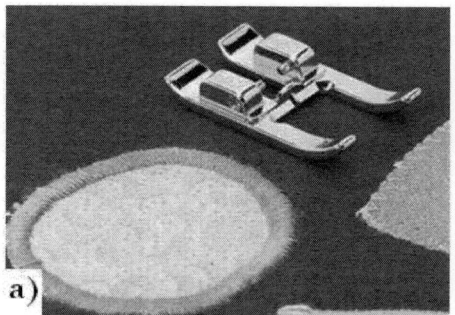 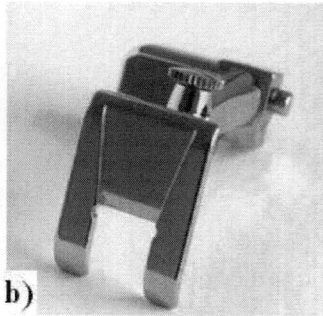

Figure 4.59   Open Toe Foot – Open Toe Appliqué Foot / Satin foot.

4.9.21 Walking foot – even feed foot

The walking foot is shown in Figure 4.60. Both the pieces of fabric are fed evenly by the feed dogs and the presser foot. This is useful when sewing fabrics that are hard to feed such as vinyl and leather and when sewing fabric that slip easily like velvets. It is good for quilting to keep all the pieces moving neatly and uniformly to avoid puckering.

This foot is unique because of its feed dog that in combination with the sewing machine's feed dog, work to move the fabric along. The fabric is actually sandwiched between the two feed dogs, and is essentially walked along by both set of feed dog.

Clamp it with needle plate screw (Helps to work synchronised with needle)

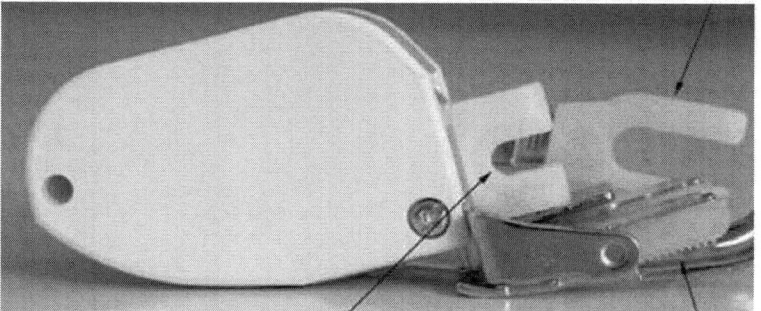

Attach with presser foot holder   **Additional feeder**

Figure 4.60   Walking Foot – Even Feed presser Foot.

- The even feed foot is a great tool for quilting and makes working on large bed sized quilts so much easier. This foot helps to prevent shifting and bunching of the quilt as it is being quilted. The foot is used for stitch-in-the-ditch quilting on the standard sewing machine.

- This foot is not just for quilting, it can also be used for regular sewing, such as top stitching of multiple layers, attaching binding to blankets and to prevent slippage of fabrics such as leather, velvet, and knits.

- The even feed foot is also one of those feet that has a standard look, regardless of the manufacturer making it easy for identification.

- The foot has an arm that attaches to the needle bar and when the needle moves up and down, one of the toes of the presser foot moves with the feed dogs of the presser foot, while the other toe picks up the next stitch. The presser foot itself attaches to the presser bar and although there may be slight variations in how the foot attaches to the machine and needle bar, the principle is the same.

4.9.22 Quarter-inch foot

- For precise 1/4 in. seam allowances, a 1/8 in. arm or marked guide can be included. A more specific quilting foot will include the 1/4 in. seam guide, but is also designed for ease in pivoting and can generally only be used for straight stitching because of a very small needle opening; the presser foot is shown in Figure 4.61.

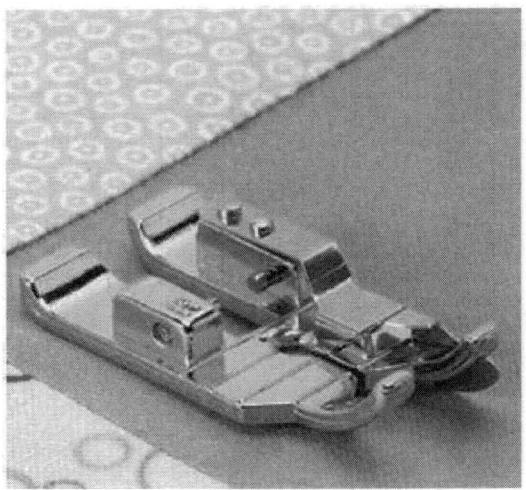

Figure 4.61   Quarter-Inch Foot.

- The foot when used to embellish can create squares, diamonds, blocks or rows that contain consistent, even stitches.

- Many of these feet instead of having a quilting guide bar that is inserted in a hole in the shank of the foot holder, actually have red markings on the foot, that can be graded in 1/4, 1/8 or 3/8 in. Other feet of this type may just have the red markings and no guide bar.

- The 1/4 in. foot made by one manufacturer has a wide opening, instead of different size toes but it does have markings on the foot. The marking on any of this foot assists the quilter not only in creating quarter inch seams but also in pivoting the quilt for the next 1/4 in. seam.

- This specialty foot can also be used for narrow seams, such as in baby garments, doll clothes or other crafting projects.

4.9.23 The fringe/looped foot

The fringe foot like so many of the specialty foot does more than just create fringe (Figure 4.62) and by having more than one function buying this foot can perhaps justify the cost of having a specialty foot that may not be used every day. This foot may also be called the looper foot by some vendors.

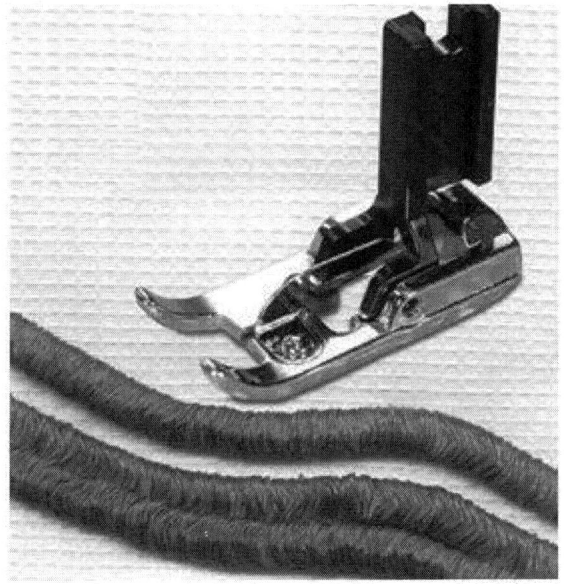

Figure 4.62   The fringe / looped stitch presser foot.

This foot can be used not only to attach fringe to garments but can also be used to attach buttons. The foot has a raised bar down its center which is vertical that is ideal for creating the shank that is needed to sew on buttons, particularly since the thread loops are formed over this bar. The stitch that is used with this foot is the zigzag stitch, this is what makes it possible to sew fringe and attach buttons.

- Fringe foot adds surface texture and applies custom trims
- Joins seams with decorative or heirloom stitches
- Creates 3D flowers by trimming threads after sewing

4.9.24 Curve master foot

The curve master foot is a very useful foot for the quilter. This ingenious presser foot is designed to sew curved quilt blocks. This foot can be used to sew quilt blocks for patterns such as drunkard's path, double wedding ring and wheel of whimsy, all of which contain circle blocks. The feature of this presser foot is that it eliminates the puckering in these blocks which can occur when trying to piece circles in the conventional way on the sewing machine.

- The foot is wide with very short toes, unlike other presser feet, which makes it easy to maneuver and join opposite curved edges. There is a 1/4 in. guide bar for stitching the required 1/4 in. seam allowance that is required for quilting. Another interesting feature of this foot is that there are multiple adaptors that allow the foot to be connected to a wide variety of sewing machine vendors; all that needs to be done is to choose the correct adaptor for the sewing machine. The adaptor is actually connected to the sewing machine before the foot is attached. The curve master comes with six adaptors and need not be bought individually.

- The curve master presser foot comes in two sizes, 1/4 in. for quilters and dolls and 5/8 in. for clothing. It comes with shafts to fit most machines, Bernina machines need an additional adapter in order for it to work. This presser foot allows to easily sew curves without pinning the pieces together. The 1/4 in. foot has a raised 1/4 in. seam allowance guide, which is used to guide the fabric against while sewing the curve. Sewing machine settings are straight forward, so there are no special needles that must be used. Universal needle can be used. A straight stitch is used and both the needle and bobbin thread are all-purpose thread. The only tool required is either a stiletto or tweezer.

4.9.25 Nylon ring foot

This is a special type of presser foot (Figure 4.64) appropriate for slippery and problem materials like velvet, plastic-coated, rubber-backed, etc. Nylon ring foot is used in single needle lockstitch machine. Nylon ring foot is not available for needle feed machine.

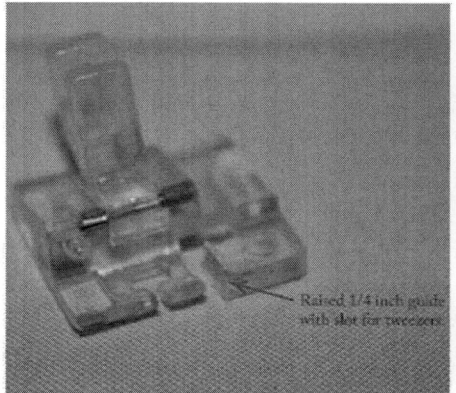

Figure 4.63   Curve master foot for curved edge sewing.

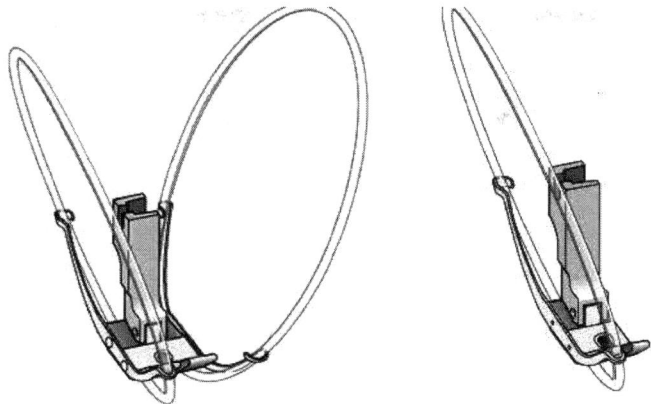

Figure 4.64   Nylon roller foot.

The single ring presser foot is very useful in the case of slippery material with small edges. The single ring on left side and narrow foot shoe enables the sewing of the small bag kind of materials.

4.10   Summary

This chapter outlines the different classification of the sewing machine attachment in two categories, based on their application and also based on their working nature. The different types of attachments coming under each category are explained with neat illustration. The applications of various attachments are discussed. The second part of the chapter explains the most important and widely used types of presser feet. The working and application of the various presser feet is explained with illustration.

Reference

1.   Solinger, Jacob, *Apparel Manufacturing Handbook-Analysis, Principles and Practice*, Columbia Boblin Media Corp., 1988.

2.   SUISEI attachments for industrial sewing machine, TOHKI Industrial Co. Ltd. Japan, 2005.

3.   http://www.bernina.com/en-US/Products-us/BERNINA-products-us/BERNINA-Accessories-us/Presser-feet-us.

4.   http://www.rga.co.uk/catalogue.shtml.

5.   http://www.jesseheap.com/index-folders-attachments.htm.

6.   http://www.atlatt.com/apparel/EQUIPMENT/folders/.

7. http://www.needlethreadhere.com/kategori/sewing-attachments.aspx.

8. http://www.sew4home.com/tips-resources/sewing-tips-tricks/how-sew-buttons-your-sewing-machine.

9. http://www.the-sewing-partner.com/.

10. http://craftygreenrabbit.wordpress.com/2013/08/19/sew-crazy-an-almost-comprehensive-concise-guide-to-sewing-feet/.

11. http://www.seasonedhomemaker.com/2013/06/sewing-machine-feet-the-edge-stitch-foot.html.

12. http://badskirt.blogspot.in/2010/07/edgestitch-foot-demystified.html.

13. http://www.donssewing.com/presserfeet/feetsewing.html.

14. http://www.acraftyfox.net/tutorial/check-out-her-curves-curve-master-presser-foot-tutorial/.

5

Sewing machine maintenance

This chapter describes the various maintenance procedures to be followed in the apparel industry with reference to sewing machines. The different types of maintenance procedures and their requirements, common preventive maintenance steps are explained and this chapter gives a clear idea about the possible sewing machine problems with their causes and troubleshooting methods.

Key words: Maintenance types, Preventive maintenance, Condition based maintenance, Breakdown maintenance, Sewing machine problems, Causes and remedies.

5.1 Sewing machine maintenance

The maintenance is defined as the act of maintaining or the state of being maintained or the work of keeping something in proper condition; upkeep. The maintenance procedure of the sewing machine can be in by either machine operator or a mechanical engineer from the apparel industry. Since the operator is working on the machine, he or she can understand the small abnormalities in the machine at the very beginning and can avoid the major issues.

5.1.1 Causes of machine malfunction

1. Improper cleaning

2. Improper tightening of machine parts during machine cleaning and adjustment

3. Improper machine part adjustment when it is set up for operation

4. Failure to replace machine parts before being worn down due to operational control limits

5. Improper lubrication

5.1.2 The sewing machine maintenance is done mainly by operator and plant mechanic

- The operator should be given a schedule for cleaning the machine.
- The time lapse between cleaning will depend how quickly lint and fabric finish substances accumulate in the stitching, feeding and other parts.
- This will depend upon fabric structure, sewing machine speed and the running time of the machine.

Examples: pile fabric causes more lint than ordinary plain fabric.

If a given fabric is on a machine operating at 6000 rpm and 90% sewing speed – more lint and finishing substance will accumulate than the same fabric being run in a machine with the speed of 3000 rpm and 40% running time.

Type of machine is also a factor – sewing machine with cutting knife or multiple needle machine will lint more than machine without such knives.

5.1.3 Operators role

- Can perform daily checking
- Can judge an abnormal condition
- If they fully understand the structure and function of the facility, then they can detect an abnormal condition easily

The operators can deal with the trouble and perform the recovery

- They can make an improvement
- They can recover (repair) the malfunctions

5.1.4 The mechanics role in maintenance is threefold

1. To teach the operator how to clean and lubricate the machine properly and to provide a check list, instruction and schedule for this.

2. To inspect the machine regularly for the worn parts which require replacement to prevent machine malfunctioning.

3. To keep the maintenance record for each machine, listing all adjustment, repairs and replacement. The record should list monthly running time, production of the machine. This is essential for determining and checking the life of the machine parts. Figure 5.1 represents a sewing machine maintenance record.

Sewing machine Maintenance Record								Card no:				
Maker:			Model No:			Serial No:				Max.RPM:		
Stitch Type:			Ded Type:			In Use:				Avg RPM:		

Monthly Running time and Production Summary:

	Jan	Feb	March	Apr	May	June	July	Aug	Sep	Oct	Nov	Dec	Year
Running Time													
Production													

SERVICE RECORD

S.No	Date	Mechnanic	Operator	Operation	Needle used	Fabric used	Time used	Part No used / repair made

Figure 5.1   Sewing machine maintenance record.

5.2   Classification of machine maintenance

The machine maintenance procedure can be classified into major categories as given in Figure 5.2.

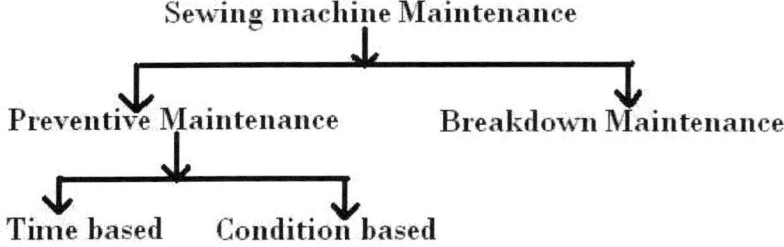

Figure 5.2   Classification of maintenance.

5.2.1 Preventive maintenance

Preventive maintenance is the process to prevent machine problems and deterioration before it occurs in a major level through the following acts:

1.   Daily maintenance to prevent deterioration (cleaning, checking, lubricating, tightening bolts)

2.   Periodic inspection and facility diagnosis

3.   Maintenance for recovery

5.2.1.1 TBM (time-based maintenance)

Time-based maintenance is the method whereby maintenance is carried out in a fixed cycle determined by the mean time between failures (MTBF) and other factors (Figure 5.3).

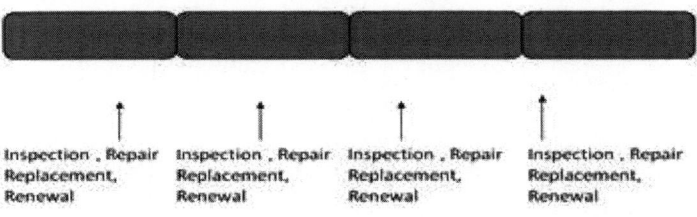

Figure 5.3   Time based Maintenance.

The time-based maintenance can be suitable for a company which runs with a higher capacity, where the machineries in different departments are continuously monitored and maintained for their condition. In this case, the expenditure of the procedure seems to be high. Hence this method was not mostly used in the small scale industries. However, one should understand that, this kind of periodical maintenance will reduce the sudden and unexpected breakdown and loss due to the process defects.

5.2.1.2 CBM (condition-based maintenance)

This procedure is also referred as prediction maintenance (Figure 5.4). This is a method whereby facility deterioration is constantly or periodically monitored by facility diagnosis and maintenance is conducted when an abnormality is found.

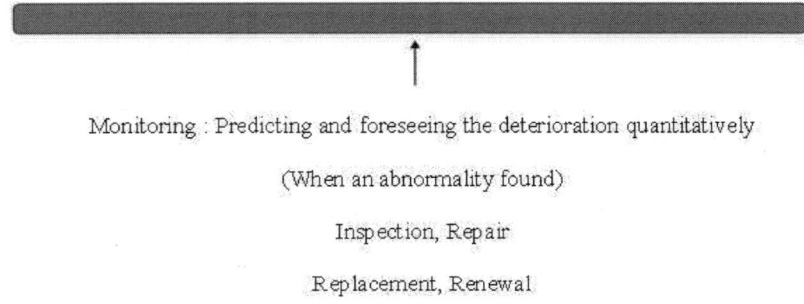

Figure 5.4   Condition based Maintenance.

This kind of procedure is generally used in most of the industries. Since, this method is cost-effective for small scale industries they prefer this kind of inspection. In this method, the machine is subjected to maintenance once the abnormality is observed by the operator or mechanical engineer. This method is preferred only for the replacements of components and where service facility is available immediately and for the kind of breakdown that may not have the greater influence on the production ability of the particular manufacturing facility.

5.2.2 Breakdown maintenance

Breakdown maintenance is carried out only when a breakdown occurs. This method is generally accepted only if the breakdown has no influence on the production ability of the particular manufacturing facility. The manufacturing units in the unorganized sectors and very small and small-scale industries prefer this kind of maintenance method. However, the industries in the organized sector also prefer this method during situations like breakdown of special purpose machines and machines which are used very uncommon. This method mostly preferred in the following cases:

- No failures frequently

- No serious effect with short time failure shutdown

- Where spare or replacement available

- Not so important

5.3   General sewing machine cleaning

If the machine begins to run hard, it is a sign that dirt or lint has jammed inside a bearing. The machine running should be continued and it should be flushed with cleaning fluid until the dirt and gummed oil are washed from the bearing. To remove any remaining dirt and oil, a cloth or brush dipped in cleaning fluid is used to scrub all parts of machine that can be reached. A needle, knife or other pointed instrument can be used to dig or scrap away any remaining gummed dirt or lint in the feed dog, around the bobbin case, and in other areas.

The lower tension of the bobbin case and the upper thread tension discs needs to be checked. The thread under the tension of the bobbin can be pulled to remove dirt. Pulling a piece of cloth soaked in cleaning fluid back and forth between the discs of the upper tension will help in removing dirt. The process needs to be repeated with a dry cloth to be sure that no lint or thread is caught between them.

In addition to general cleaning, three areas need special attention. They include:

- The hook and bobbin areas assembly
- The needle-bar and presser foot
- The handwheel bearing and the clutch assembly

5.3.1 Bobbin and hook area

Lint is the primary offender in this area. The bobbin case can be removed on all makes of machines. A dry brush can be used to clean out all lint. Any thread that may be wound up around the hook shaft should be removed. On many machines, the hook assembly can also be removed for more complete cleaning. Placing one drop of oil on the exterior perimeter of the hook and the bobbin race lubricates it after cleaning. The bobbin and hook area are shown in Figure 5.5.

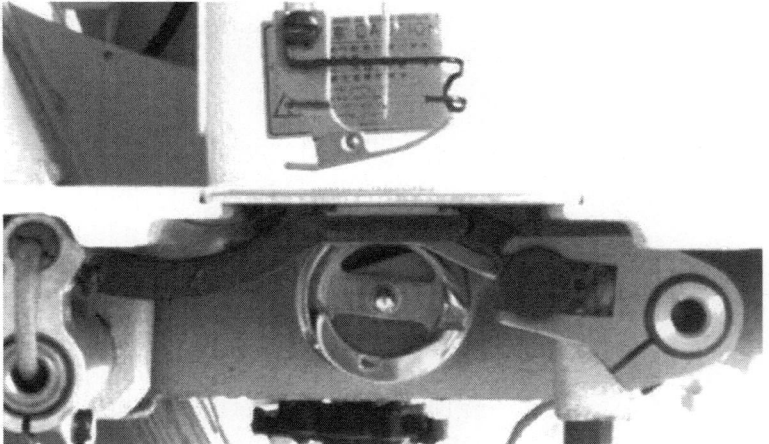

Figure 5.5   Bobbin and hook shuttle area of single needle lock stitch machine for maintenance.

5.3.2 Face plate area

The face plate on most machines is held in place with one or two screws. By removing these, the plate can be easily removed for cleaning of the needle-bar and presser foot bar. On some of the newer machines, the face plate is a part of a housing that is mounted on hinges, which makes it easy to move the entire housing away from the bars and mechanisms behind it. No other parts need to be removed for cleaning in this area. First a dry brush can be used to clean out all lint and other foreign material. A small piece of cloth with a little solvent on it can be used to clean the needle-bar and presser bar of any gummy grease.

After thorough cleaning, a drop or two drops of oil can be put on each shaft where it slides through the housing. All the other moving parts should be oiled according to the instruction book before replacing face plate, as shown in Figure 5.6.

Figure 5.6   Face plate of a single needle lock stitch machine.

5.3.3 Hand wheel area

To remove the clutch and hand wheel, the small screw should be loosened in the face of the locknut, the handwheel of a sewing machine is shown in Figure 5.7. Next, the locknut must be unscrewed and the washer and hand wheel should be removed. The position of the washer should be noticed so that it can be put back in the same position. The hand wheel should slide off the shaft easily. If the machine is driven by an external belt, this belt will have to be removed before the hand wheel comes off. Then the hand wheel, washer and shaft should be cleaned. The shaft should be lubricated with two drops of oil and a small amount of grease should be placed on all gears. The hand wheel and clutch should be reassembled. If the clutch fails to operate, either because it is not able to hold or fails to release, the locknut should be removed again and the washer should be turned one half turn (180°) and reassembled. The clutch should then work properly.

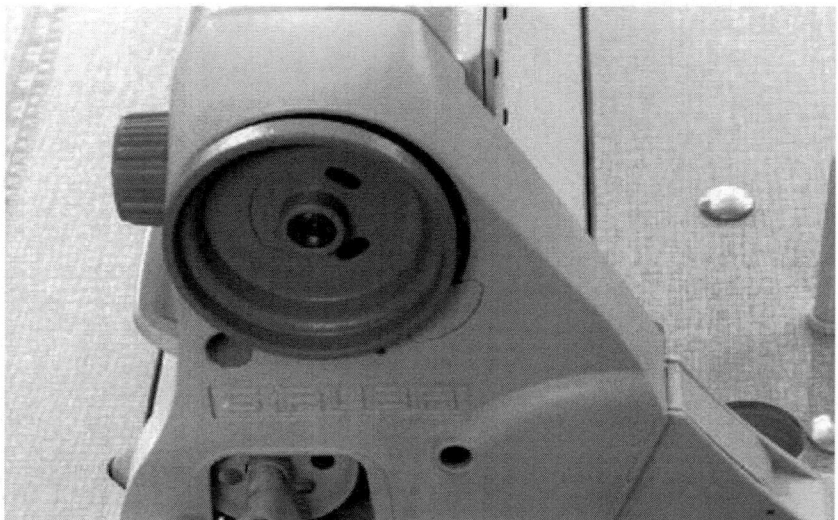

Figure 5.7   Hand wheel area of a single needle lock stitch machine.

5.4   Preventive maintenance of sewing machine

5.4.1 Cover the machine always

- Dust, lint, grit and animal hair can find their way into the machine and cause all sorts of problems, especially for the printed circuit board of a computerized machine.

- So it should not be placed near an open window and it should always be covered when not in use.

- A ready-made plastic cover can be purchased from a notion or machine dealer.

- The machine should be kept under wraps when sewing is not done (Figure 5.8).

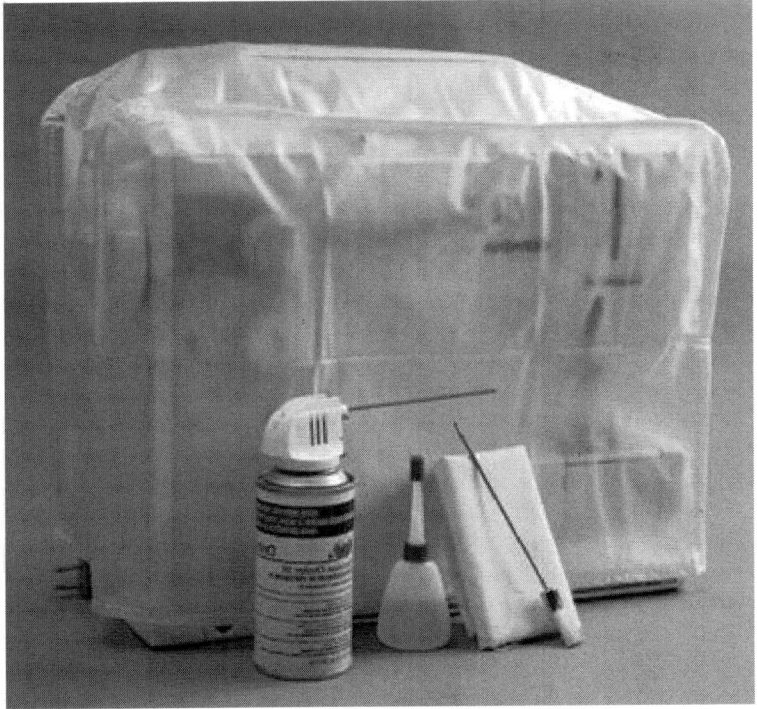

Figure 5.8 Sewing machine covered properly.

5.4.2 Frequent needle change

Replacing the needle after every 8 h of sewing time is essential based on the application. When sewing is done, the needle passes through the fabric thousands of times per minute, and each time it does two things:

1. It makes a hole in the fabric for the thread to glide through

2. It forms a loop with the thread to make the actual stitch

The bobbin hook picks up this loop by moving just 0.05 mm or less behind the needle, it is about the thickness of a piece of paper, so if the needle becomes bent or dull, it may result in skipped stitches, broken or looped threads, runs and pulls in the fabric or even damage the machine.

• When the needle is compatible with the fabric and thread, the machine sews more smoothly.

• An inappropriate needle will force the thread through the fabric instead of letting it glide cleanly through the needle hole and may cause broken or sheared threads.

- A common mistake is to use a needle that is too small for the thread. A sharp needle, like a microtex or jeans needle is the better choice when sewing natural-fibre woven fabrics than the universal needle, which has a slight rounded tip that gently pushes the fibres aside as it enters the fabric. Microtex needle inserts itself precisely in the fabric where it is pointed.

- Regular ballpoint needles, however, are still the best for sewing knits, fleece fabrics and elastic. Figure 5.9 shows the damaged needles.

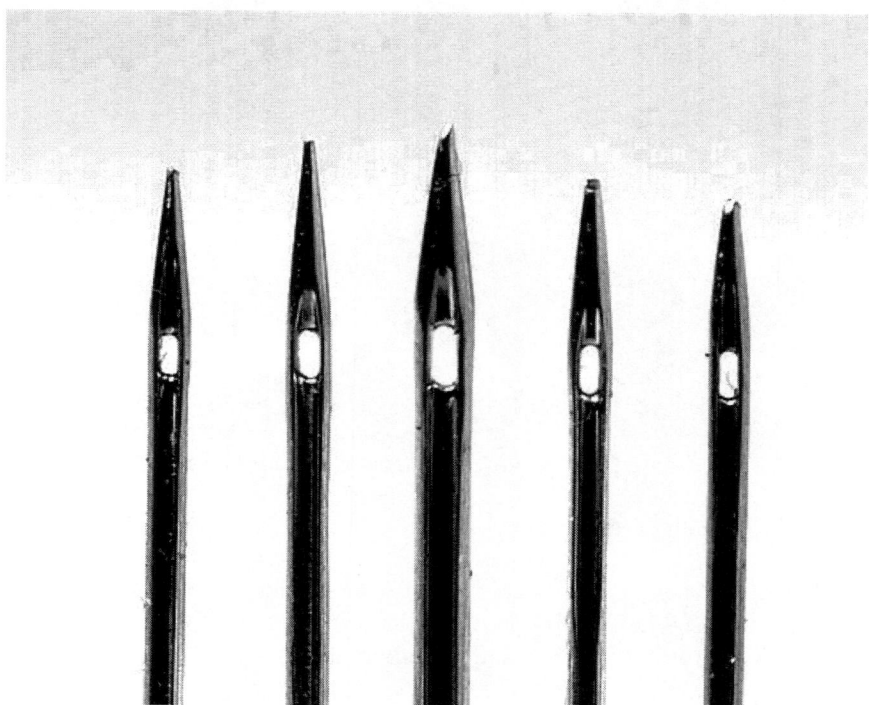

Figure 5.9   Damaged sewing machine needles.

5.4.3 Wind bobbins correctly

- It should be made sure that there are no thread tails hanging from the bobbin when it is inserted into the bobbin case. They can jam the machine and cause the upper thread to break.

- It should be understood that there is no such thing as a generic bobbin, every machine has its own type of bobbin, an example of bobbin is shown in Figure 5.10.

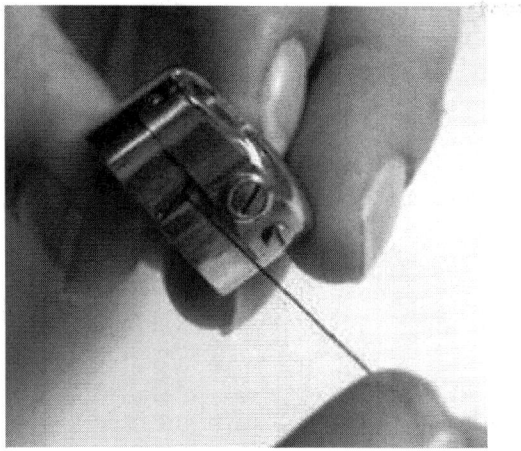

Figure 5.10   Single needle lock stitch machine bobbin.

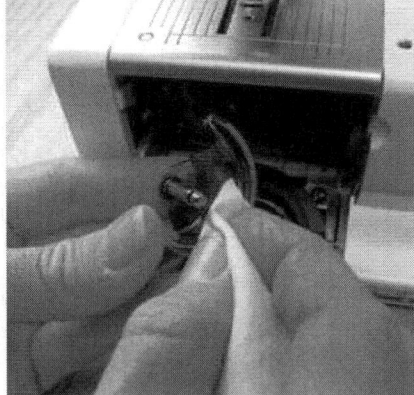

Figure 5.11   Cleaning the Bobbin and bobbin case.

- Always a bobbin designed for the specific machine should be used in order to avoid skipped stitches, loose threads and noise, as well as permanent damage to the bobbin case.

- It is important to clean inside the bobbin case. Compressed air with a straw should be used to direct air to a desired spot, blow out lint and loose threads as shown in Figure 5.11.

- A soft piece of muslin should be used with a very tiny droplet of sewing machine oil to clean the race hook. If the hook is removable, a drop of oil can be put on it before returning it to the machine.

5.4.4 Cleaning of tension disc

A piece of muslin can be folded in half and used to clean between tension discs.

- Starting from the top, the tension discs should be cleaned with a folded piece of fine muslin (Figure 5.12).

- It should be ensured that the presser foot is up, so that the tension springs are loose and the muslin can move easily between the discs, dislodging any lint or fuzz.

- Using a can of compressed air, blow air from back to front, to remove loose particles from around the tension discs and to clean other areas inside the machine.

- Blowing into the machine by an individual should be avoided as the breath contains moisture and will eventually cause corrosion.

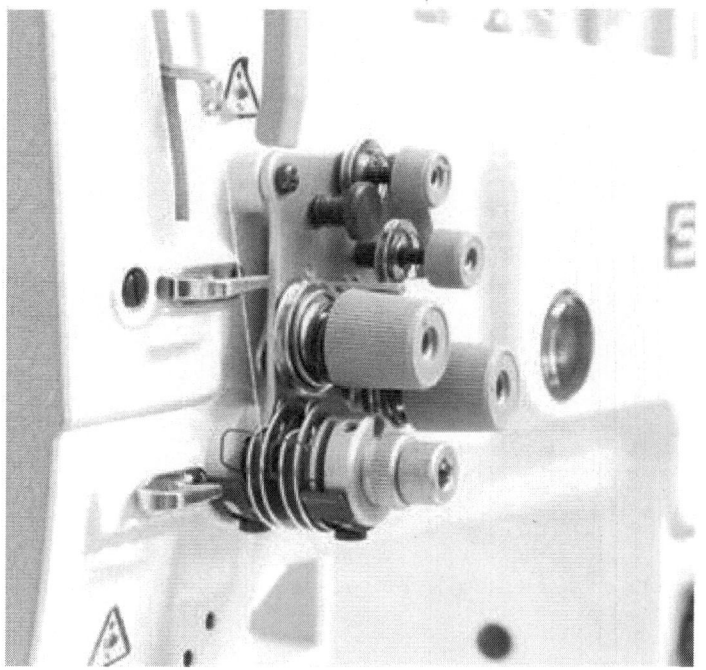

Figure 5.12   Sewing machine tension discks.

5.4.5 Oiling-sewing machine

- It should be made sure that all the dirt from inside, including the pieces of thread stuck inside the machine must be removed. Lubrication is an important aspect of sewing machine repair and maintenance (Figure 5.13).

- Grease is applied on the gears and oil to the other parts of the sewing machine.

- White lithium grease is ideal to keep the sewing machine well lubricated. One drop of oil on each bearing and in each oil hole will be sufficient.

- It is a good practice to oil the machine after each day's work or after 8–10 h of use.

- If the machine requires a lubricant, the suggested areas can be lubricated. The lubricant recommended for the machine should be used for best results. After oiling and lubricating the machine, the excess oil should be wiped and the machine should be kept in its position.

Figure 5.13   Oiling the sewing machine parts.

5.4.6 Drive belt monitoring

- Drive belt of the sewing machine is one of the most important components of the sewing machine.

- Irregular maintenance or overuse of the machine can either make the belt loose or wear out, and thus affect the performance of the sewing machine.

- If the belt has loosened due to frequent use, it has to be tightened, and if it has worn out, proper care should be given and have to replace it.

- Covers of the machine can be opened using a screwdriver and the drive belt should be checked to verify whether it needs to be tightened or replaced. The belt should not flex more than half an inch, if it does, it means the belt is loose and the necessary steps given below should be followed to tighten it.

- If it is noted that the belt has cracked or depreciated considerably, then it should be replaced and a trial run should be taken after reassembling the machine. There are different types of belts used in various sewing machines and some of them are given in Figure 5.14.

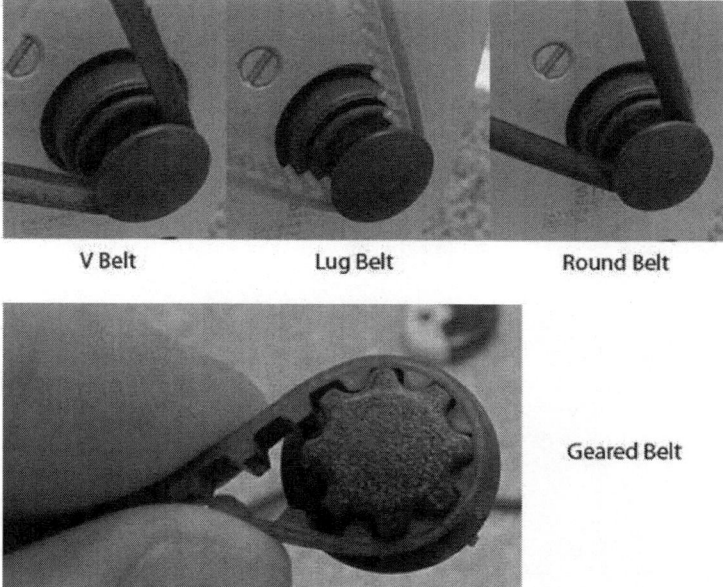

| V Belt | Lug Belt | Round Belt |

Geared Belt

Figure 5.14   Sewing machine belt types.

1. Lug belts
2. V belts
3. Round belts
4. Geared belts

V belts

This type of belt is almost identical to the lug belt, except there are no 'teeth' inside of this belt.

Lug belts

This type of belt is by far the most common belt used for sewing machines and can be mistaken for a geared belt because most of them have 'teeth' on them that resemble a geared belt. The difference between this type of belt and a geared belt is these 'teeth' do not fit inside indentations on the motor pulley.

Round belts

This belt is used over a lug belt or V belt because it stretches, so if the measurement was off, it will still work. They grip better and provide better traction with less slipping.

Geared belts

If the machine uses a geared belt, the exact belt made for the specific model should be used. A geared belt has 'teeth' like the lug belt, but these 'teeth' fit inside grooves on the motor pulley.

5.4.7 Breaking of needles

- The needle size must be proper for type of machine used and also for selected fabric and thread.
- Presser foot must be securely fastened to the bar and that it helps the needle go through the centre of the presser foot.
- During the sewing process, the fabric should not be pulled. This may cause the needle to become bent, resulting in the needle striking the metal throat plate. Keeping the fabric 'taut' is alright, as this does not put added stress on the needle.
- The needle should be tightly fastened and all the way up into the needle bar.
- Sewing over pins and metal zippers/parts should be avoided.

5.4.8 Upper thread breaks

- If the needle is backward or not set in completely then the upper thread will break all the time. (The flat side of the needle is generally opposite to the last thread guide above the needle and the direction of threading).
- The secondary reason is improper threading. It should be made sure that thread is not caught in a place where it is not supposed to be or catching on rough places. This will create too much of tension for needle thread.
- Needle is too small for selected thread or blunt or bent needle would cause this breakage.
- It should be made sure to place spool on machine in a way that the thread feeds off the spool smoothly and that the edges of the thread spool are smooth.

- Thread may not be of good quality. Buying of bargain thread; it ends up causing more lint to be deposited in your machine and may break more frequently. Threads may even have knots that will not go through the needle.

5.4.9 Lower thread breaking

- The first reason for the breaking of lower thread is improper threading of the bobbin case which is similar to the breaking of the upper thread. This will cause too much of tension on the thread and causes breaks.

- Thread on bobbin overfilled or not wound evenly.

- Too much tension on thread during bobbin winding.

- Poor sewing thread quality and this causes fur generation and frequent breaks along with the knotted thread in the bobbin.

- Burr on edge of throat plate (caused by needle hitting it).

5.4.10 Feed system maintenance

- The needle plates should be in good condition, the damaged needle plates will affect the feeding of fabric either by creating improper feed or causes needle break by hitting the needle.

- The needle hole in the needle plate should be approximately 30% larger than the needle size. This may be affected by needle hit on needle plate. It is necessary to make sure that the needle hole size/ needle size relationship is correct at every situation. This avoids major problems in feeding.

- Check to make sure the needle hole in the needle plate is not damaged.

- Look for needle damage and sharp edges and check to see if the needle plate is flat and not bent down at the needle hole.

Feed dogs

- Feed dogs should be in good condition and correct for the application, the damages in the feed dog must be checked and removed.

- The pitch length of the feed dog is basically the teeth per inch. This has to be selected properly based on the type of fabric used for the sewing. Generally the following should be followed:

  - For lightweight fabric sewing – 20–24 teeth per inch

  - For medium weight fabric sewing – 14–18 teeth per inch

- For heavyweight fabric sewing – 10–12 teeth per inch

Presser foot

- **The presser foot should be of correct size and type** and also the **needle hole size/needle size relation should be correct** to have sufficient clearance during sewing.

- The presser foot must have proper pressure on the front and back of the needle.

- When the **foot is flat on the needle plate**, there should not be any space, it should not be possible to insert a thin piece of paper between the foot and the needle plate from the front or back side of the needle. This will help the friction between the fabric and presser foot and makes sure that the fabric is fed correctly.

- There should not be any excessive wear on the fabric by pressure. The pressure on the presser foot should **be as light as possible** and gives a uniform stitch length. It should be checked to make sure that the correct stitch length is being used.

5.5 Final checks before running machine

5.5.1 Checking to see if the machines are being kept clean

- Machines should be blown off every day to remove lint and trash.

- On lockstitch machines, the hook should be blown off regularly during the day to prevent lint or dirt from building up in the oil ports in the race of the hook.

5.5.2 Checking to see that the machines are being lubricated regularly

- Oil levels should be checked daily and additional oil should be added if necessary

- Randomly oil levels should be checked in the machines

- A high quality white machine oil should be used that will not stain

- The availability of proper machine oil in the factory should be checked

- It should be made sure that the oil is not contaminated

- It should be checked to see that oil reservoir pump filters are cleaned regularly

5.5.3 If compressed air is used, it should be made sure that the air system is regulated properly and has humidity dryers, filters and lubricator in the air lines.

It should be checked for rusted areas due to excessive moisture in production area.

5.5.4 Checking machines for wear on critical moving parts and screw

- It should be checked for shake in needle bar due to worn needle bar bushing
- Excessive movement in stitch forming devices, etc. should be checked
- It should be checked if there are any missing screws
- Defective screws that are difficult to tighten properly should be identified, if any

5.5.5 Checking condition of mechanics tools to see that they are being maintained properly

Checking to make sure the correct thread type and sizes are being used

- If thread vendor is specified, order book and inventory should be checked to make sure that the proper thread is being used. Thread stands should be in proper condition and the top eyelets should be oriented properly.
- Optimum distance between the top of cone and top eyelet should be there (should be no more than one cone higher than cone size being used). Thread stand eyelets should be smooth and not grooved or damaged.
- Cone should be held in a vertical stationary position. Machine eyelets and guides should be smooth and not grooved, rusted or damaged. Machine thread tensions should be as light as possible but should still give balanced stitches. The tension post must be observed to see how much of it is exposed beyond the tension nut.

- Generally the more of the tension post that is visible, it means that more tension is being applied to the sewing thread. The condition of the tension discs, take-up spring, etc. should be observed.

5.5.6 Checking the bottom tension to make sure it is as loose as possible and still give a balance stitch in lockstitch machines

- Bobbin winder should be in good condition and should be making correct wind on bobbins. Condition of bobbins and bobbin tension should be checked.

- The nicks on edge of bobbins indicating incorrect needle height should be looked for, it should be checked for any damaged or bent bobbins. Bobbin tension (minimum tension recommended) must be checked.

5.6 Other systems

5.6.1 Material trimming systems

It is necessary to see that the cutting knives are sharp and are trimming properly. This can be checked by placing a single end of thread between the knives and observing if the thread is cut by the knives.

5.6.2 Thread trimming system

It is mandatory to see that the chain cutters are cutting the chain to the proper length that minimizes trim.

5.6.3 Folders and guides

It should be ensured that the correct capacity folder is used for the fabric being sewn. It should be checked for additional folds which would be caused by excessive folder capacity. Any excessive stretching of the fabric should be avoided as it causes additional seam puckering.

5.7 Trouble shooting sewing problems

Some of the major sewing machine problems with their causes and remedies are provided for better understanding in Table 5.1.

Table 5.1   Sewing machine problems with their causes and remedies

Symptom	Root Cause	Corrective Action
• Sewing machine suddenly stops during sewing. • Light is on, and hand wheel can be turned easily by hand.	• Sewing machine has been run at a low speed for an extended period of time. • To prevent overheating, motor power is automatically turned off.	• Turn power off and wait about 20 minutes. Safety device will reset, and machine will be ready to operate.
• Needle will not move.	• Upper thread has run out. • Presser foot is up. • Bobbin winder shaft was left in winding position. • Buttonhole lever was not lowered when machine was placed in buttonhole mode.	• Replace empty spool and thread machine. • Lower presser foot. • Move bobbin winder shaft back fully to left. • Lower buttonhole lever.
• Sewing machine will not run.	• Presser foot is not the correct one and needle hits presser foot. • Needle has come out and is in hook of machine.	• Replace presser foot with new one. • Remove needle and insert new one.
• Upper thread breaks.	• Threading is not correct. • Thread has a knot in it. • Thread tension is too tight. • Needle is bent or blunt or has a sharp eye.	• Correctly thread machine. • Remove knot. • Correct thread tension. • Replace needle.

- Needle is wrong size.
- Needle has been inserted wrong.
- Needle and thread does not match.
- Starting to stitch too fast.
- Thread take-up lever has not been threaded.
- Misaligned off winding from thread package.
- Trapping at package base.
- Thread trapped at thread guide.
- Snarling before tension disc.
- Excessive tension.
- Broken check spring.
- Sharp edges on throat plate, hook point, needle guard, bobbin case, needle groove or eye.
- Thread fraying at needle.

- Replace needle with correct size.
- Properly insert needle.
- Start machine at a medium speed.
- Use proper thread or needle.
- Check threading order.
- Ensure that the overhead guide is directly above cop stand pin, at 2½ times the height of the thread package. Use a foam pad to prevent package tilting.
- Reduce the thread stand height to prevent vibration and spillage. Use a foam pad to prevent trapping after spillage.
- Can occur after thread breaks. Rethread correctly.
- Increase the wraps on pre-tension thread guides and reduce disc tension. Ensure discs are smooth.
- Use a stronger thread or adjust tension.
- Replace and adjust.
- Polish rough edges and replace if necessary. Replace the needle being used with a higher quality needle.
- CUse finer thread or coarser needle, as appropriate.

	• Excessive needle heat; groove or eye blocked with melted fabric. • Hook overheating. • Poor quality thread.	• Improve the fabric finish. Change to a better needle, style and finish. Apply needle lubricant via thread. Use a needle cooler. • Ensure adequate oil supply. Check the needle to hook clearance. • Change to a correctly finished thread of better quality.
• Bobbin thread breaks.	• Bobbin has not been fully inserted in bobbin case. • Bobbin has been incorrectly threaded. • Bobbin does not turn smoothly in bobbin case. • Lint in bobbin case or shuttle. • Badly wound thread on the bobbin. • Tension too tight or bobbin over-running. • Sharp edges on bobbin case or spring or looper eyelet.	• Securely install bobbin into bobbin case. • Correctly thread bobbin case. • Check to see that bobbin has been wound evenly. • Clean bobbin case and shuttle. • Adjust bobbin winder alignment. Use pre-wound bobbins. • Adjust bobbin case tension. Insert a washer or a spring to prevent over-running. • Polish edges and correct surfaces.

• Skipped stitches.	• Thread tension is too tight.	• Correct thread tension.
	• Needle is bent or blunt.	• Replace needle.
	• Needle is wrong size.	• Replace needle with correct size.
	• Needle and thread does not match.	• Use proper thread or needle.
	• Thread take-up lever has not been threaded.	• Check threading order.
	• Light pressure on presser foot.	• Increase pressure on presser foot.
	• Incorrect setting of needle.	• Reset needle.
	• Hook, looper or needle failing to enter thread loops at the correct time.	• Check machine clearances and timings. Check if the needle is inserted and aligned correctly. Use a needle with a deeper scarf.
	• Thread loop failure caused by incorrect needle size / style for thread size / type.	• Change needle size / style.
	• Thread loop failure due to incorrect setting of thread control mechanism causing thread loop starvation.	• Reset to standard and check loop formation with a strobe.
	• Flagging of fabric due to poor presser foot control or too large a throat plate hole.	• Re-adjust the presser foot pressure. Change the throat plate to match the needle.
	• Needle deflections or bent needle.	• Use a reinforced needle, reset the needle guard and replace the needle.

	• Incorrect sewing tension in the needle or under threads. • Poor thread loop formation.	• Re-adjust the tensions. • Check with a strobe. Change to superior spun polyester or filament based corespun threads.
• Stitches are not formed properly.	• Thread has not been pulled into thread sensor guide. • Threading is not correct. • Bobbin case has been threaded wrong. • Spool cap is wrong size for thread spool.	• Fully pull thread into thread sensor guide. • Correct threading. • Correctly thread bobbin case. • Replace the cap with correct size.
• Thread Fusing when the machine stops	• Poorly finished or incorrect thread. • Densely woven fabric that is poorly or harshly finished. • Damaged or overheated needle after thread breakage.	• Use better quality thread. • Improve fabric finish. Change to more suitable needles. Apply needle coolants. • Change the needle.
• Irregular stitches.	• Incorrect size needle. • Improper threading. • Loose upper thread tension. • Pulling fabric. • Light pressure on presser foot. • Loose presser foot. • Unevenly wound bobbin.	• Choose correct size needle for thread and fabric. • Rethread machine. • Tighten upper thread tension. • Do not pull fabric; guide it gently. • Increase pressure on presser foot. • Reset presser foot. • Rewind bobbin.

• Imbalanced / Variable Stitching	• Incorrect sewing tensions.	• Check for snarling, adjust thread tensions.
	• Incorrect threading.	• Rethread machine.
	• Needle thread snagging on bobbin case or positioning finger.	• Polish bobbin case surfaces. Reset positioning finger and opening finger.
	• Variable tension due to poor thread lubrication.	• Switch to superior quality threads
• Staggered Stitching	• Needle vibration or deflection.	• Increase needle size or change to a reinforced or tapered needle.
	• Incorrect or blunt needle point.	• Change the needle.
	• Incorrect needle-to-thread size relationship.	• Change needle or thread size as appropriate.
	• Feed dog sway.	• Tighten the feed dog.
	• Poor fabric control, presser foot bounce.	• Reset the presser foot. Change the feed mechanism
	• The pressure of the Presser Foot is insufficient. The fabric is not held down firmly.	• Increase the pressure of the Presser Foot
	• The Presser Foot is not firmly attached to the Presser Bar.	• Tighten firmly the screw that holds the Presser Foot to the Presser Bar.
	• The Feed Dog is not firmly attached to the Feed Bar.	• Tighten firmly the screws that hold the Feed Dog to the Feed Bar.
	• The Feed Dog is not adjusted straight in the machine.	• Loosen somewhat the screws that hold the Feed Dog to the Feed Bar, and then adjust the Feed Dog so that it is in line with the direction of sewing. Tighten the above screws firmly after this adjustment

- The Thread Take-up Spring (Check Spring) may be defective. It is also possible that it is not adjusted correctly and that its tension is too weak.
- The loop of the thread take-up spring is bent sideways.
- The color of upper and lower thread, being different from the color of the fabric, will make staggered stitches more obvious.

- IReplace a defective spring with a new one.
- Adjust the movement and the tension of the spring. The upward movement of the loop of the spring must have stopped at the moment the Needle is about to enter the fabric.
- Bend back the loop of this spring to its correct shape. If badly defective, replace the spring with a new one.

- Fabric puckers.

- Stitch length is too long for material.
- Needle point is blunt.
- Incorrect thread tension.
- Light pressure on presser foot.
- Fabric is too sheer or soft.
- Using two different sizes or kinds of thread.
- Variable differential fabric feed.
- Incorrect thread balance.
- Improper thread type.

- Decrease stitch length.
- Replace needle.
- Reset thread tension.
- Increase pressure on presser foot.
- Use underlay of tissue paper.
- Upper thread and bobbin thread should be the same size and kind.
- Improve the fabric feed mechanism. Replace worn out feed dogs. Reduce the maximum sewing speed.
- Ensure proper balance between the top and bottom thread.
- Use threads with controlled elongation. Properly maintain tension guides

• Bunching of thread.	• Upper and lower threads not drawn back under presser foot. • Feed dog down.	• Draw both threads back under presser foot. • Raise feed dog.
• Needle breaks.	• A thin needle was used for sewing a heavy weight material. • Needle has not been fully inserted into needle bar. • Needle clamp screw is loose. • Presser foot is not correct one. • Presser foot is loose. • Pulling on fabric as you sew.	• Use correct size needle. • Properly insert needle. • Securely tighten needle clamp screw. • Use correct presser foot. • Reset presser foot. • Do not pull fabric, guide it gently.
• Loud noise is heard. • Knocking noise, machine jammed.	• Dust has accumulated in feed dogs. • Lint is in hook. • Thread caught in shuttle.	• Clean machine. • Clean machine. • Disassemble machine and clean shuttle.
• The machine does not feed material.	• The stitch length has been set to zero. • The presser foot pressure is too low. • Feed dogs are lowered. • Threads are knotted under fabric.	• Reselect the proper stitch pattern and length. • Set presser foot pressure adjustment lever to "normal". • Raise feed dogs. • Place both threads back under presser foot before beginning to stitch.

• Threading cannot be done. • The needle threader will not turn. • The threading hook will not enter needle eye.	• The needle is not in highest position. • The needle threader is designed not to turn to protect itself if the needle is not up. • The needle has not been fully inserted into needle bar.	• Turn handwheel until needle reaches its highest position. • Properly insert needle.
• Needle threader cannot be returned and the sewing machine stops.	• Sewing machine was accidentally started while threading hook was still in needle eye (during threading).	• Slightly turn handle clockwise and remove threader.

5.8   Other sewing defects due to machine problems

Figure 5.15   Damaged throat plate due to improper sewing.

5.8.1 Looped stitches

- The main cause for looped stitch (Figure 5.16) formations is incorrect threading in the machine. Every time it has to be made sure that thread passes correctly through the thread take up and upper tension disk(s).

- Tension disc(s) and guides must be cleaned after every shift, because the dirt accumulated in it will prevent the smooth passage of the thread.

- The bobbin may not be threaded correctly. It should be ensured that the thread passes through the tension slot correctly. In machines that have a removable bobbin case, the bobbin should turn in a clockwise direction when the thread is pulled. Mostly drop in bobbins will turn counter clockwise.

- Correct size of the needles should be used for the selected thread.

- Bobbin housing may be nicked. In some machines with plastic parts, it is not uncommon for them to become nicked by the needle. Nicks should be smoothened with emery cloth or housing should be replaced with a metal housing available from the dealer.

Figure 5.16   Looped stitch formation.

5.8.2 Loop or knot on top surface of fabric

If the upper thread lies straight on the top surface of the fabric and the lower thread appears there in form of small knots, then the possible causes may be:

- The tension of the upper thread is too tight or the tension of the lower thread is too weak.

- The tension spring of the bobbin case is bent out of its proper shape. This condition does not provide sufficient tension of the lower thread.

- Lint, dirt or pieces of thread might have accumulated underneath the tension spring of the bobbin case. Rusty or rough spots might be there between the tension discs.

The remedies are:

- The tension of the upper thread should be loosened or the tension of the lower thread should be tightened, until the tensions of both threads are correctly balanced.

- If the spring is deformed, then it should be carefully shaped with a small round-nosed plier. If not successful, the defective spring should be replaced with a new one.

- The tension spring should be removed, cleaned and then replaced correctly.

- Bobbin case should be cleaned properly.

- Any rusty or rough spots should be removed with fine emery cloth and polished with crocus cloth. Badly defective tension discs should be replaced with new ones.

5.8.3 Loops or knots on the underside of the fabric

If the lower thread lies straight on the underside of fabric and the upper thread appears there in form of loops or small knots, then the reasons may be:

- The tension of the upper thread is too weak or the tension of the lower thread is too tight. There could be dirt, lint or pieces of thread between the tension discs.

- The upper thread might have cut deep grooves into the tension discs.

- The bobbin case might not be threaded correctly. The bobbin might unwind itself in the wrong direction.

- The head of the tension adjusting screw of the bobbin case might be protruding excessively and hence it might catch and retard the upper thread.

- The stitch hole in the throat plate or feed dog might be too small.

- The bobbin might not be evenly wound. It might unwind itself irregularly.

- The bobbin might be damaged or bent and might not revolve freely and evenly.

- The point of the needle might be bent over ('hooked point').

- Lint or dirt in the bobbin case might prevent the bobbin from revolving freely.

- The needle might not be correctly timed in relation to the loop taker (hook).

The possible remedies are:

- The tension of the upper thread should be tightened or the tension of the lower thread should be loosened, until the tensions of both threads are correctly balanced.

- The upper tension disc should be taken apart, then cleaned and replaced.

- The defective tension discs should be replaced with new ones.

- The bobbin case should be threaded correctly.

- The screw should be adjusted correctly. If the condition continues to persist, the head of the screw should be smoothened with fine emery cloth or the defective screw should be replaced with a new one.

- The throat plate or feed dog should be replaced with one that has a larger stitch hole.

- The defective needle should be replaced with a new one.

- The bobbin should be replaced with one that is wound evenly.

- With a pointed pin, the dirt from the part should be removed and then should be cleaned with kerosene. Then it should be dried thoroughly before replacing in the machine.

It should be made certain that the needle is pushed all the way up into the needle bar and then held there firmly. The point of the loop taker should be adjusted in relation to the needle. For 'timing' follow the general rule: The needle must have risen about 3/32 of an inch from its lowest position at the moment the point of the loop taker is just at the centre of the rising needle. At that moment the point of the loop taker must be about 1/16 of an inch above the eye of the needle.

5.8.4 Skipped stitches

- One of the main reasons for skipped stitches is fabric finishing. If the fabric was not prewashed, it will cause skipped stitches. If the finish cannot be removed by prewashing, then a thread lubricant should be applied to the needle (Figure 5.17).

- Wrong size or type of needle for thread/fabric may also be a reason. Needle may be bent or blunt or the needle may be incorrectly set into needle bar.

- A ball point or stretch needle can be used for sewing knit fabric and a sharp needle can be used for woven fabric.

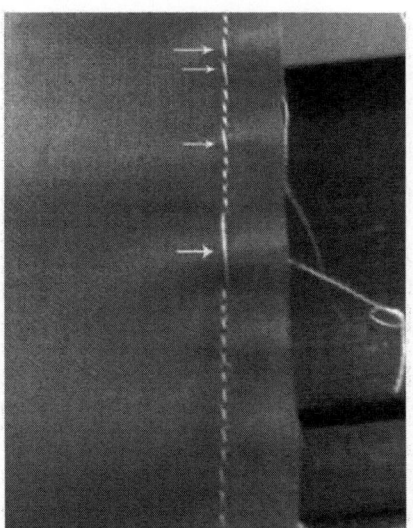

Figure 5.17   Skipped stitches.

5.8.5 Fabric puckering

5.8.5.1 Classification of puckering

The different types of sewing puckering defects were classified in Table 5.2.

5.8.5.2 Fabric feed puckering

It occurs due to poorly controlled fabric feed by sewing machine feed mechanism. When two plies of material are not fed uniformly, the variations are held captive by the stitches and cause feed pucker.

It generally occurs:

- If the foot pressure on the machine is too high, excessive friction can stretch the top ply. If the foot pressure is too low, the foot can bounce momentarily losing control of both plies.

Table 5.2 Classification of sewing puckering defects

Pucker Type	Test to Identify	Solutions
Tension pucker: Incorrect tension settings	Cut all top and bottom stitches without distorting the thread loops If pucker disappears, the cause is incorrect tension settings	Sew with minimum tension Adjust feed timing for maximum pull off Use smaller thread Loosen bottom thread tension Select thread with good lubrication
Inherent pucker: Structural jamming	Cut and remove all top and bottom stitches If pucker disappears, the cause is structural jamming	Sew on bias if possible Use smaller needle size Reduce stitch density Change stitch type
Pucker caused by fabric and thread instability	Apply a visual check – Does the pucker appear after treatment or washing?	Use synthetic threads with low wet shrinkage Compatible garment components
Feed pucker: Poorly controlled fabric feed	Make two cuts across the seam in areas of maximum pucker. Remove all stitches between the cuts to see if one ply is longer than the other	Minimum foot pressure Correct the feed dog Correct operator technique Match feed and foot Correct hole size in throat and pressure foot Correct top and bottom feed timing Low friction presser foot

Figure 5.18   Feed puckering during sewing process.

- When the operator stretches one ply more than the other as they are fed into the machine.

- Many seams exhibit both conditions when the operator attempts to correct the unequal feeding of the fabric into the machine.

- To find out if pucker is due to unequal feeding, then make two cuts across the seam and then remove all sewing threads between the cuts. If one ply is longer than the other, then the pucker has been caused by unequal feeding.

To avoid feed puckering, a low friction presser foot should be used or presser foot should be adjusted for optimum pressure and also the positive and even feed of the fabric must not be affected.

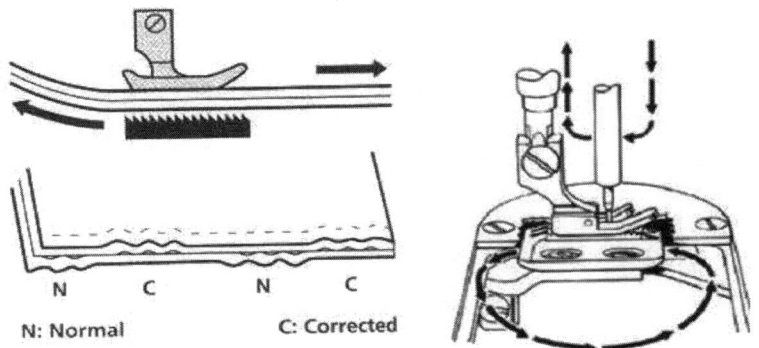

Figure 5.19   Formation of puckering during feeding process.

- The back of the feed dog can be raised slightly to create a pulling effect away from the needle.

- The feed dog should be checked for correct height, teeth per inch and number of rows of teeth which are appropriate for the particular fabric and operation.

- It should be ensured that the operator is not holding back on either the top or bottom ply and it should be checked for fabric hanging in any folders that may be in use.

- The throat plate and presser foot should have needle holes approximately twice the size of the needle.

- If the machine has both top and bottom feeders, it should be ensured that the timing is correct.

5.8.5.3 Tension pucker

Tension pucker is caused while sewing with too much tension, thereby causing a stretch in the thread. After sewing, the thread relaxes. As it attempts to recover its original length, it gathers up the seam, causing the pucker, which cannot be immediately seen; and may be noticeable at a later stage. Tension is also referred to as seam shrinkage or thread relaxation pucker.

To identify this, both the top and bottom threads of all stitches should be carefully cut along a few centimetre of the seam, without distorting the thread loops in the fabric. If the pucker is relieved over this length, then it should have been caused by thread tensions.

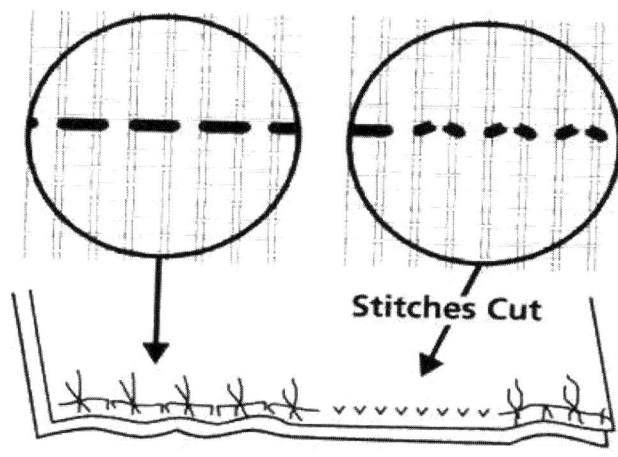

Figure 5.20   Tension puckering during sewing.

Remedies

- The tension applied to the thread should be reduced while it is being wound onto the lock stitch bobbin.

- The needle thread tension should be set as light as possible while achieving a balanced stitch at the same time. This will reduce the amount of puckering while the thread is stretched and improve the sewability.

- A high-quality sewing thread should be used with a low-friction lubricant applied to it. This will allow the thread to run smoothly through the thread guides and tension controls.

- A high-quality sewing thread should be used with even unwinding tension for smooth flow of thread to the sewing area.

- It should be ensured that the sewing machine feed timing is correctly set, as incorrect feed timing can lead to the need of applying excessive tension to the needle thread. Incorrect timing may lead to an imbalanced stitch.

5.8.5.4 Shrinkage pucker

Dimensional changes in thread or fabrics during post-sewing treatments or washing can cause pucker, primarily because threads and fabrics react to these processes differently. For example, soft cotton threads increase in diameter and shorten in length when wet, as they absorb moisture. This can distort the fabric. Even though the thread may return to nearly original dimensions when dry, the fabric can remain puckered.

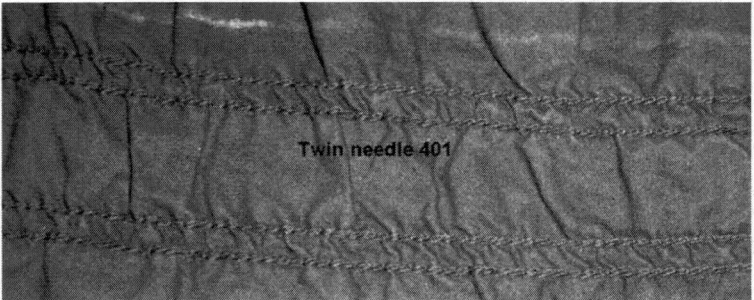

Figure 5.21   Shrinkage pucker during sewing.

To identify this, as such, there is no accurate scientific method to determine this type of pucker. However, a trained eye can find the pucker with a simple visual check – the pucker will appear on the garment after treatment/finish or washing.

Remedies

The best way to avoid this kind of pucker is by using synthetic threads with low wet shrinkage properties. Also it should be ensured that all components of the garment are compatible. For example, if a lining or reinforcing tape shrinks more than the base fabric, the base fabric will pucker along the stitch line.

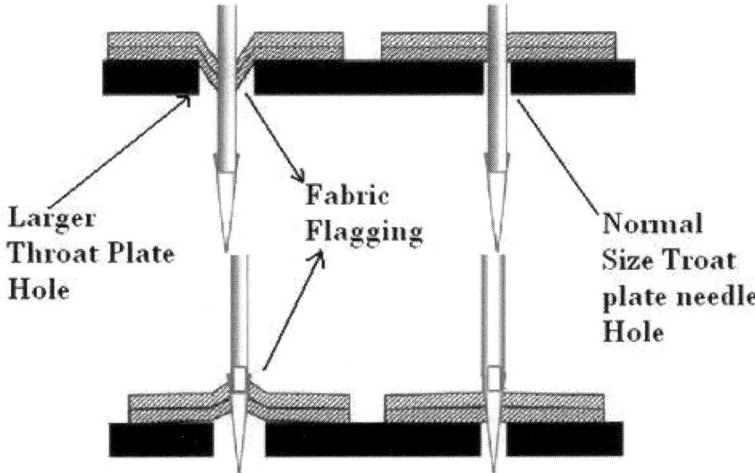

Figure 5.22   Fabric flagging defects in cloth due to damaged throat plate.

Similarly, if two pieces of material with different extensions are stretched during sewing, different relaxations can cause seam pucker on one face of the composite.

5.8.6 Fabric flagging

Choking of fabric inside the needle hole on the needle/throat plate is called fabric flagging in sewing process. This creates fabric damages in sewed materials, the possible causes for the flagging issues are:

- The throat plate aperture enlarges due to wear and tear.

- While sewing, the needle pushes the fabric through the aperture before penetrating the fabric.

- This can also happen when the needle size (thickness) is changed and if the throat plate is not changed accordingly.

To avoid this problem, the following remedies must be taken

- Throat plates must be changed at regular intervals after checking for wear and tear.

- Throat plates must be changed in accordance with the needle size even if there are no signs of wear and tear.

5.9 Summary

The importance of maintenance is clearly spelt out in this chapter. The classification of various types of maintenance is explained with examples. More thrust is given to the maintenance of sewing machinery and role of operators in maintaining proper condition of sewing machines are also discussed. Detailed information about maintenance of all parts of sewing machine is provided. The importance of preventive maintenance and various preventive maintenance activities carried out in sewing department is explained. This chapter also discusses the troubleshooting procedures for various faults occurring in the apparel machinery. The causes and remedies for those faults are detailed elaborately. More emphasis is given to the causes and remedies of fabric puckering and fabric flagging. The chapter covers the general maintenance and troubleshooting activities carried out for apparel machinery.

References

1. Solinger, Jacob, *Apparel Manufacturing Handbook-Analysis, Principles and Practice*, Columbia Boblin Media Corp., 1988.

2. Carr, H. and Latham, B., *The Technology of Clothing Manufacture*, 2nd edn., Blackwell Scientific, Oxford, 1994.

3. http://www.sewingsupport.com/sewing-machine-troubleshooting-guide-title-page.html.

4. http://www.threadsmagazine.com/.

5. http://www.chfmagazine.com/.

6. http://sewandserge.com/tshoot.asp.

7.   http://aces.nmsu.edu/pubs/.

8. http://www.coatsindustrial.com/en/information-hub/apparel-expertise/solutions-to-sewing-problems.

Finishing machineries

This chapter deals with the various pressing and fusing machineries used in the apparel industry. The commonly used pressing machineries like buck press, steam air finisher for different applications, various trouser pressing machineries and steam tunnel are discussed based on their working and application. The various principles used in the fusing process are also detailed along with the fusing machine classification and their quality parameters.

Keywords: Buck press, Steam air finishers, Trouser press, Steam tunnel, Continues fusing, Radio frequency fusing, Quality parameters.

6.1 Finishing machineries

Garment finishing machines are an integral part of garment making. These machines play the role of making garments in a presentable form free from creases and crushing. Pressing operation plays a major part in creating finished garments. The aesthetic appeal of the garment is increased with pressing. In the garment manufacturing process, the fabric is subjected to lot of stress during cutting, sewing process. Due to the processes, crease formation and crushing of garments take place and it happens during handling and transportation also. The look of a garment is increased manifold times due to proper pressing of garments thereby attracting the public to buy the garment. We will see the function of pressing and machines used for it in detail.

6.2 Functions of pressing

The functions of pressing are

1. To remove the unwanted creases and crush marks

2. To induce creases to the garment based on the design requirement

3. To enable the garment fit to the contour of the body

4. To enable further sewing by preparing the garment for next sewing operation

5.   To finally finish the garment for packing

(1) To remove the unwanted creases and crush marks – During manufacture of garments, creases form in the garment and also during the process, garments are subjected to crushing due to its movement from place to place and handling of the operator. The creases and crushes formed give a bad appearance to the garment and hence they need to be removed. Pressing does the work of removing those creases and crushes. More creasing will be seen in the case of garments handled in bundles by tying up tightly or when piled on trolleys and boxes.

(2) To induce creases to the garment based on the design requirement – In trousers, skirts and shirts, creases are required at certain places to make it look attractive. With the pressing operation, creases can be formed at required places like shirt collars, pleated skirts, edges of hem and cuffs, pocket flaps, waistband top edge's and patch pocket edges.

(3) To enable the garment fit to the contour of the body – Pressing helps in shrinking and stretching. Pressing is done to make the garment fit to the contour of the body especially in wool fabrics. It is called as moulding. After moulding it is not possible to unpick the seams and return garment parts to their former flat state.

(4) To enable further sewing by preparing the garment for next sewing operation – There are two terms used in pressing – Under pressing and final pressing (Top off). Under pressing is the pressing operation on semi-constructed garments to make the garment parts ready for further sewing. Final pressing is the process of pressing the completed garments finally before packing the garment. Under pressing operation helps the garment parts to be pressed for making further sewing easier.

(5) To finally finish the garment for packing – During under pressing, the surface of the fabric would have changed temporarily because the various garment parts are under pressed for easy sewing. It sometimes results in gloss or glazing caused by extreme pressure of press in order to achieve a firm edge or seam. In order to remove the effects of under pressing and to make the garment ready for packing, final pressing operation is carried out and the garments are pressed to get a finished garment.

6.3   Classification of garments based on pressing

The garments can be classified based on the amount of pressing required. They can be classified as garments requiring no pressing, minimal pressing required garments, garments requiring use of an iron in underpressing and final pressing, garments requiring extensive underpressing and final pressing and garments requiring permanent press or pleating.

6.4 Ways of pressing

Pressing is carried out using various means. Pressing is done by application of heat, moisture and pressure. Moisture is used usually as steam. Various pressing equipments are available which use steam for pressing.

6.5 Pressing equipment and methods

In this topic, let us see about the use of various pressing equipment and methods. The various equipment used are iron, steam presses, steam air finisher, steam tunnel, pleating and permanent press.

6.5.1 Iron

Iron which is known commonly as iron box has been in existence for a very long time with the traditional one being used with heat generated using charcoal. There were also irons which were heated by gas flame inside the metal casting and it was mainly used in touching up of men's jackets. However in recent times, steam and electric irons are used commonly, as shown in Figure 6.1. The iron is heated by using an electric element. The heating is controlled by thermostat and steam is supplied. The weight of iron ranges from 2 to 15 **kg**. In industries, ironing is done at a range of workplaces. Simple pressing tables similar to domestic ironing board are used for ironing pants and shapes of garments. Nowadays tables have supply of vacuum to hold the garments to the base for ease of drying and setting after ironing.

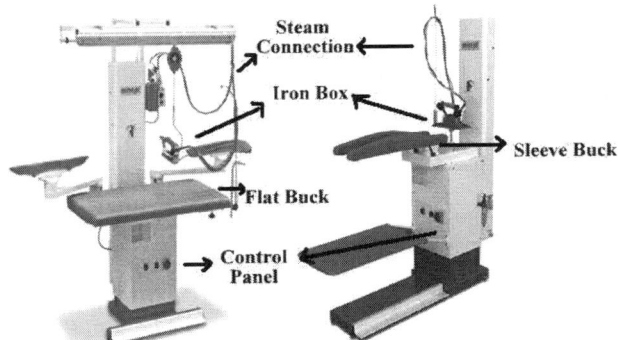

Figure 6.1   Ironing units with varieties in shapes.

6.5.2 Steam press/buck press

Steam presses commonly consist of a static buck and a head of complementary shape closing onto it, thereby sandwiching the garment to be pressed. It consists of a frame housing the buck which is normally in round shape for pressing different garments and linkages to close the head by a scissor action. Steam is passed to head and buck using a pipe system. Adequate controls are provided for controlling head closure and vacuum. Vacuum is created to provide suction through the buck using a vacuum system. The typical pressing cycle is as follows:

A garment need to be pressed is fixed in the buck

↓

The buck head closes and locks

↓

Then steam is applied to the head or the buck to press the garment for a predetermined time

↓

The buck head is released

↓

Vacuum is applied to the garment to cool and dry it

↓

The garment then moved around the buck for the next part of it to be pressed

Similar operations are carried out for completing the pressing of the garment and then the garment is hung on the hanger. It should be ensured that the garment is not damp or distorted after pressing. Bucks are available commonly for all the garments with universal shapes called universal bucks as shown in Figure 6.2. Bucks are used when large quantities of a garment shape are being manufactured. Even manually operated scissor action presses have been improved considerably by the use of electronically controlled pneumatic power. Various types of presses are used in steam pressing are shown in Figure 6.3. The different steam presses used for different parts of the garments are collar press, sleeve press, shoulder press, back and front press, collar master.

6.5.3 Carousel press

Carousel press is a new development in pressing operation in which a pair of bucks is provided that rotates between operator and the head. The head can be of single or double based on the bucks being identical or an opposite pair for pressing the left and right of a garment part.

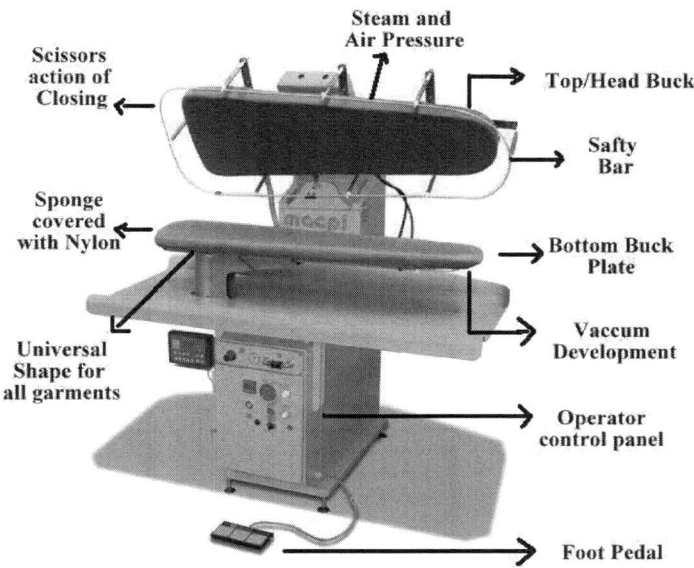

Figure 6.2   Parts of Universal buck Press.

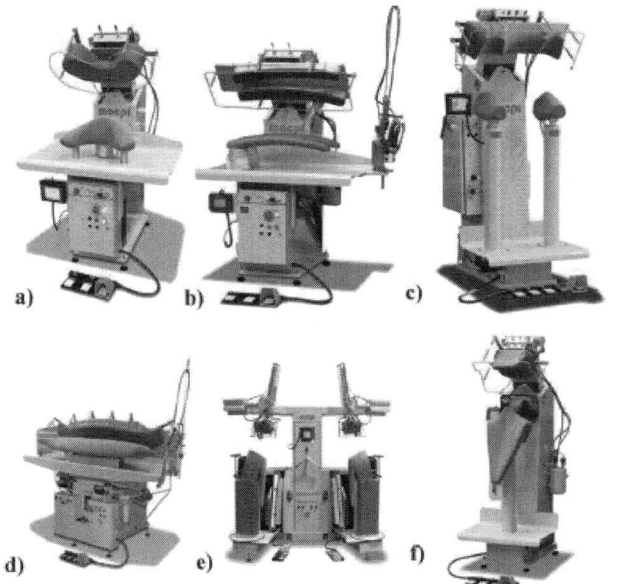

Figure 6.3 Different Steam/ Buck presses (a) Collar Press (b) Sleeve Press
(c) Shoulder Press (d) Back and Front Press (e) Shoulder sleeve
press (f) Collar master.

In this press, scissor action and vertically acting heads can be used. The operator is kept away from the steam in this press as he/she loads the garment onto one buck which is then moved away to be aligned under the head behind the screen thereby preventing the operator from steam. The other buck can be loaded when the machine carries out the controlled pressing cycle there by resulting in higher output. The typical carousel press is shown in Figure 6.4.

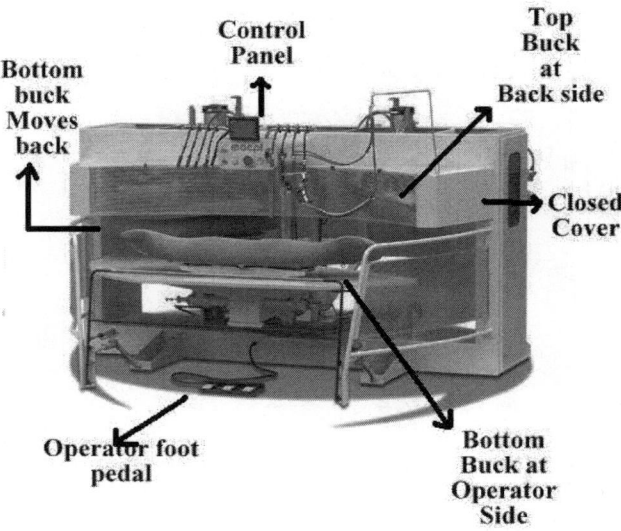

Figure 6.4   Carousel Press for High production.

6.5.4 Trouser pressing

Trousers include a wide variety of garments, ranging from jeans, women's trousers with simpler construction and requiring a less sharp crease, men's trousers including four pockets, and suit trousers (Figure 6.5). The trouser pressing is carried out in two operations along with underpressing of the seam. The first operation is done for legging on a flat press to set and crease the legs and the second operation for topping in a series of lays around the top of the trouser on a contoured press. The above method being a conventional one, a number of machines have been introduced where topping and legging are combined into one operation. It is done by vertically suspending the trousers over an upright buck from the waist band and then pressed from each side. The problems by using this method are the maintaining consistency throughout the size range and flexibility of mechanical pressure on critical areas such as waist bands, pockets and fly. The process of combining topping and legging depends on the construction of the trousers, crease sharpness and quality of the trouser required. So separate legging and topping are also used as per the requirement.

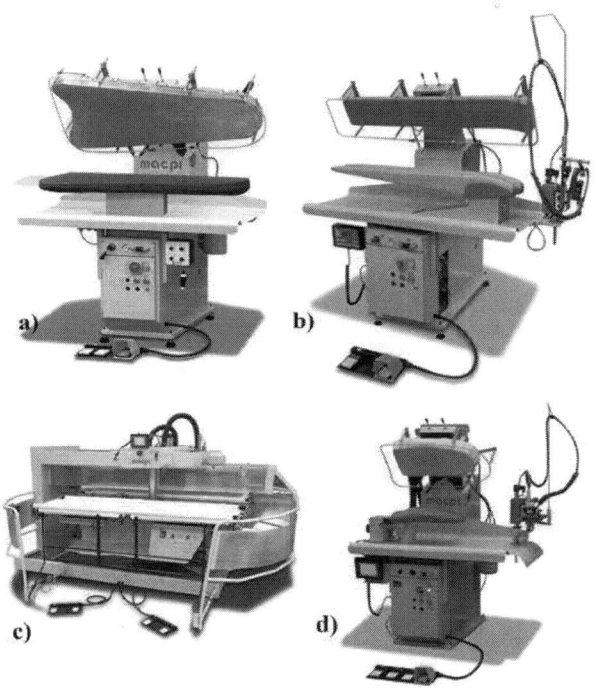

Figure 6.5 (a) Single legger for trouser press (b) Side seam finisher (c) Carousel for trouser (d) Topper.

6.5.5 Double legger-pressing machine

Double legger-pressing machine is used for pressing trousers, and in this machine both the legs are pressed simultaneously with the top hanging down between two separate bucks. These machines consist of vertically acting heads, carousels and microprocessor controls. Heat resistant silicone foam is used for covering bucks of steam presses and tables used with irons and vacuum boards and the outside being covered with a woven polyester cover. Sometimes stretch nylon is used with the highly contoured bucks.

A split head section applies steam on the seam area of the full length of the leg portion to avoid the impression and gloss at the seam side. The underside of the silicone foam is protected from continuous heat of the buck by a layer of synthetic felt. Even distribution of steam is ensured by covering the heads of steam press with a combination of materials. Various materials are arranged in layers with metal head outside followed by metal gauze layer, synthetic felt layer, main layer of knitted cotton padding and a final layer of outer cover as on the buck. Figure 6.6 shows a double legger-pressing machine.

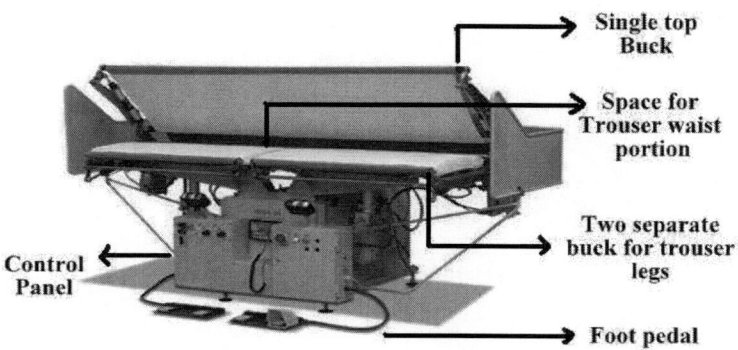

Figure 6.6   Double legger for trouser pressing operation.

6.5.6 Creasing machines

Creasing machines are used to fold over and press the edges of clothing components such as pockets or cuffs to prepare them for easy sewing. Blades are used to create creases and folds. The component is kept over a die with blades aiding in forming creases around it and required pressure is exerted during pressing cycle. Creasing machine is actually preparation-underpressing machine. Creasing machine principles and components are basically alike regardless of type or style of the section being creased by the machine.

Most creasing machines employ steam to help make and stabilize the crease. Some fabrics do not require steam for edge creasing, where the machine can be operated without steam application also. The heat in such machines was furnished by electrical heating element. These machines are made on three basic operation models: manual, semi-automatic and automatic. Creasing machine with automatic timing device permits one to get the highest degree of pressing quality with maximum efficiency in labour, capitalization, space and utility cost. The working cycle of an automatic creasing machine is very similar to the buck pressing or form pressing operation as follows:

1. Pick up and position the section on the press bed

2. Crease press (includes the situation of machine, manual or automatic)

3. Extract – Manual or automatic

4. Discard – Manual or automatic

These machines (Figure 6.7) crease all types of pockets for shirts, blouses, working clothes and trousers. Electro-pneumatic workstation with two loading and creasing stations, each independently controlled. The machine is therefore used for overlapping operation with only one operator. Handling time is considerably reduced and quality is improved compared to sewing on of pockets without edge creasing and to edge creasing with irons. Two different pocket types can be pressed in overlapping operation. Adjustable to the full range of placket sizes. Built-in lamp makes it easier to position stripes and patterns. A sword ensures proper creased edges and also includes vacuum pump for base.

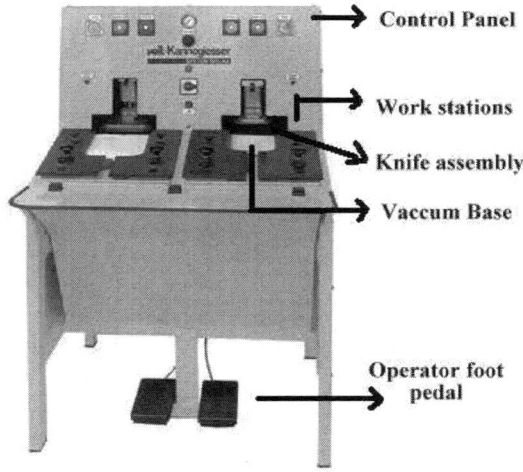

Figure 6.7 creasing machine for apparel finishing.

6.5.7 Steam air finisher/form pressing machine/dolly press

Form presses are used largely for off pressing operations, the final pressing operates on the manufactured apparel product before it is packaged and distributed to the retailer. All the form presses used in the apparel manufacturing are based on the principle of expanding a collapsible bag, made in approximate form of the finished product, with steam or air. Rigid form presses such as the metal sleeve forms are comparatively rare in use. The exact construction detail of an expansion bag form press varies with the type of product for which the press has been designed.

Regardless of the product, all form presses consist the following components: machine frame, steam distribution system, air distribution system, the form bag with size and style adjustment, gauge for steam and air pressure, actuation for steam and air release, regulation for steam and air pressure and timer for steam and air cycle.

Steam air finishers have separate steam and compressed air distribution system housed in a frame and a canvas bag pressing form in the approximate shape of the garment to be pressed. These finishers are also known as a 'puffer', a form press or a 'dolly' press. It is called as a finisher instead of a presser due to very less pressure exertion during pressing and the entire garment is pressed at the same time thus reducing the time in positioning and repositioning the garment. These finishers are fitted with controls for steam and air release. Timers are used for controlling the steam and air cycles. The garment is pressed by putting it on the form from above and then expanding the form to its full size and shape with steam being blown through it from inside. A cycle of 8 s steaming is followed by hot drying and also by blowing from inside. Steam air finisher removes accidental creases and it cannot be used for forming creases. It is useful for pressing daily wears like night gowns, T-shirts, etc. However, there will be a need for pressing hem separately with a flat press or steam press. Figure 6.8 shows the steam air finisher for apparel finishing.

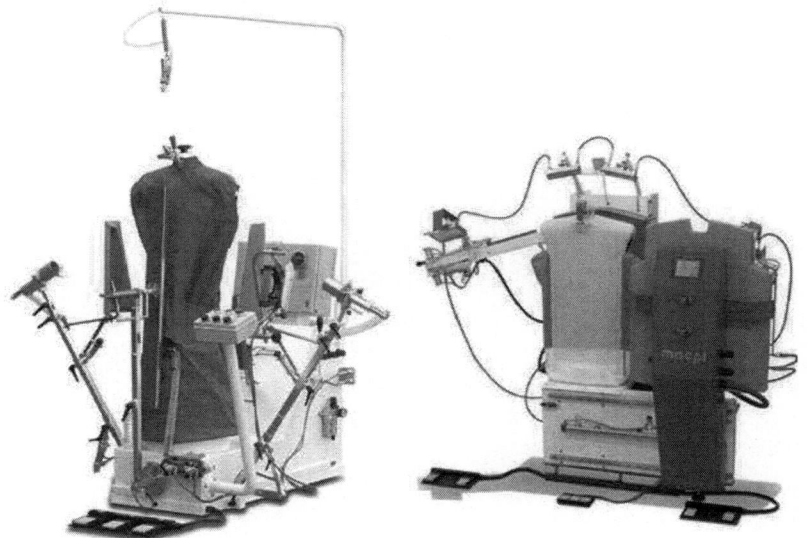

Figure 6.8   Stream Air finisher for apparel.

6.5.8 Steam tunnel

Steam tunnels are used for finishing knitted goods. They can be used for man-made fibre garments and their blends also. This garment finishing process involves no pressure application and reduced handling of garments in steam tunnel (Figure 6.9). In this finishing process, the garments are put on hangers and fed through a cabinet using a motorized rail. The garments pass through sections with superheated steam and it is dried by blowing air. In some cases garments are loaded onto frames and passed through the tunnel on a conveyor. Steam helps to relax the fibres in the garment and the tunnel helps in avoiding the need for any other pressing process before or after this operation. In some cases, it completely eliminates the other pressing processes. These tunnels are incorporated with infrared drying in some cases. As the garments are vertically hung, the turbulence of blown air provides additional energy to remove wrinkles in woven fabrics. Proper care should be taken during pressing operation for fibres where excessive agitation causes fabric deformation.

Figure 6.9   Stream Tunnel finisher for Synthetic apparel.

6.5.9 Pleating

Pleating is the process of creating pleats in the garment. Pleats are a type of fold actually formed during stitching by doubling fabric upon itself and securing it in place. However, these pleats can also be introduced in pressing by creating a set of creases in the garment and making it set by pressing. The pleats can even be according to a geometrical pattern. Pleating is done by using pressure, moisture and heat. There are two types in machine pleating. One is a blade machine in which pleats are formed by the action of blades and then set by heat and pressure when they pass through a pair of rollers and the other type is a rotary machine in which the rollers are fitted with complimentary dies. These dies will be similar to the gears and they create the pleats when the material passes through (Figure 6.10).

Figure 6.10   Fabric pleating machine (die) for apparel.

This machine is capable of various pleating combinations and enables you to have your ideal pleating patterns. The pleat can be used for blouses, shirts, jacket cuffs and collar, school uniforms, antimacassars, bedclothes, ribbon, curtains, carves and products made of leather.

The standard working width is 1600 mm. This machine is the professional equipment mainly applied to automatic pleating and heat setting of various chemical fabrics and blended fabrics. It is capable of different pleat patterns, such as knife pleat, "I" pleat, bamboo leaf pleat, wave pleat, and the combination of various pleat, which are subject to arbitrary combination and changes. With the changes of the die cutter, varieties of patterns can be made. The processed fabric is applicable in fashion clothing, boutique clothing and professional suit. It also widely used in textile industry.

6.5.10 Permanent press

Permanent press is mainly used for crease recovery after washing for cellulose fabrics. The usage of permanent press involves treating the fabric with a post-curing resin during manufacturing followed by drying and then making garments out of it. However, the usage of permanent press gradually reduced with more polyester/cotton and polyester/viscose blends coming up as polyester helps in crease recovery and as known to all, it imparts good strength also. Once the garments are ready, they are passed through the ovens where the resins in the garment are cured at high temperature. Then the garments are washed and dried and it results in the fabric returning to the required shape. Permanent presses are used in the case of trousers as there will be requirement to form creases in the seams, hems and pockets.

6.6   Fusing machineries

The process of bonding the interlining to the outer fabric using a thermoplastic resin is called fusing. There will be base cloth in the interlining consisting of thermoplastic adhesive resin on its surface that will melt when heated to a specific temperature. The limitation of fusing is that some garments cannot be fused.

The advantages of fusible interlinings are as follows: reduced manufacturing time, less labour cost, replaces complex operations in large area applications such as jacket front, low skilled labour can be used, replaces sewing in small area applications such as reinforcing tapes, gives opportunity for alternate garment making methods and it is easy to achieve consistent quality.

6.6.1 Requirements of fusing

The requirements of fusing are given below:

- Fusing of interlinings may affect the air permeability and crease recovery. Crease recovery depends on the grain of the fabric. Proper care should be taken while selecting interlinings.

- Aesthetic qualities like draping, softness, handle required in the finished garment should be showed by the laminate produced by fusing.

- The bond strength of the laminate should be sufficient to withstand handling during garment manufacturing and the bond must resist temperature, agitation of washing and dry cleaning.

- The bond should be complete and delamination should not occur as these will appear as bubbling on the outer fabric if they occur.

- There should be no thermal shrinkage in the outer fabric due to fusing.

- There should be no strike through or strike back occurring during fusing as strike back affects the equipment parts and strike through might show as a pattern of dots of resin on the right side of the garment. When the adhesive resin is pressed into the garment fabric, it should not go right through the face side and that it does not go to the back of the interlining base cloth.

- There should not be any thermal shrinkage in the outer fabric due to fusing as fusing occurs at 150°C and fabrics may be subjected to thermal shrinkage.

- The heat of fusing may cause dye sublimation and fabrics may change colour to a level that is unacceptable, causing a mismatch between the fused and unfused parts.

- Since the fusing process involves pressure, the pile fabrics may be subject to CRUSHING which shows dissimilarity between fused and unfused parts with respect to appearance.

- In shower proof fabrics, there is a possibility that the presence of a fused interlining may WICK water and hence water resistant interlinings have been developed.

6.7   The fusing process

- Fusing process is carried out by applying temperature and pressure over a period of time. The electric heating elements of the press cause the rise in temperature at the glue line which is the interface of resin and outer fabric. The resin is active in that area.

- The resin changes from a dry solid state to a viscous fluid. There should be adequate pressure applied to regulate the flow of resin among the fibres of outer fabric and the fusible base cloth. On cooling, the resin re-solidifies and forms a bond between the two components of a laminate.

- Heat has to pass through the fabric to activate the resin and every fusible resin has an optimum temperature.

- If optimum temperature is not maintained, even higher pressure or longer time will not compensate and the bond will become weak.

- If temperature is too high, it will result in strike through or strike back.

Apart from the outer fabric, three factors determine the properties of the fused laminate:

- Base fabric of the interlining

- Type of fusible resin

- Pattern of application of the resin to the base cloth

6.7.1 Base fabric of the interlining

- Base fabrics are available in woven and non-woven constructions as sew-in interlinings and warp knits.

- Warp knits may be of lock-knit or weft insertion construction.

- Lock-knit with more yarns and more interlinking of loops is stable than simple warp knit structure.

- Nylon is mostly used due to its soft handle and excellent draping.

- Weft insert fabrics consist of vertical chains of loops with yarns laid horizontally and interlaced with the vertical chains.

6.7.2 Type of fusible resin

Various types of resins are available to cover a range of laminating requirements. The requirements of resins are:

- Fusing temperature must not be high as it will damage the outer fabric or its colour. The usual temperature is 150°C and the maximum temperature is 175°C.

- The temperature for fusing should not be lower than 110°C. If it is low, then the bond will be of inadequate strength to withstand garment making. However, leather may require low temperature.

- The resin must provide a bond that is suitably resistant to washing and dry cleaning.

- The thermoplastic nature of the resin must be such that adjustment of temperature is sufficient to permit it to penetrate the outer fabric to give a bond without strike through or strike back. It must contribute to the desired handle of the laminate.

- The resin that is used should be white or transparent, harmless and must have a low dye retention process. The resin used should not be affected by higher temperatures. The resin used should have high degree of cross-linking ability.

The types of resins available are:

- **Polyethylene**: Available in different melt flow index. Melt flow index estimates the extent to which the resin flows during operation. If the flow is easy, then the resistance of the laminate to the solvents will be lower. Washable resins are also available with higher density and they can be dry cleaned. They are used in interlinings of shirt collars.

- **Polypropylene:** This resin is similar to high density polyethylene but it reaches its softening point at a higher temperature. It is suitable where rapid, high temperature drying is a part of laundering. It can withstand temperatures up to 150°C.

- **Polyamides:** Various types of nylons are employed to vary the melting range and lower the softening temperature. They are used in dry cleanable garments. Usage of polyamide resins result in washing at higher temperature and dry cleaning can be done at lower temperatures.

- **Polyesters**: They are used in garments that are washable and dry cleanable. They are less water absorbent and washing resistant.

- **Polyvinyl chloride (PVC)**: They are printed on to base fabrics as a plasticized paste. These resins are both dry cleanable and washable and are used in large areas such as coat fronts.

- **Plasticized polyvinyl acetate (PVA)**: These resins are available in the form of a continuous coating for fusing leather and fur at low pressure and temperature. These resins are not dry cleanable and have limited washability.

6.7.3 Methods of applying resins to base cloths

- Scatter coating – Particle size of resin should be in a range between 150 μm and 400 μm

- Dry dot printing – Particle size of resin should be in a range between 80 μm and 200 μm

- Paste coating – Particle size of resin should be in a range between 0 μm and 80 μm

Scatter coating

- It employs specifically designed scatter heads to provide even scatter under automatic control.

- Resin is softened in oven, pressed on base cloth and then cooled.

- It is an inexpensive method of making a fusible but product is neither uniform nor flexible as printed coatings.

Dry dot printed coating

- In this type of coating, powdered resin is filled onto engraved holes on a roller.

- The base cloth is passed over a heated roller and against the engraved roller. The powdered resin adheres to cloth in the form of dots.

- Oven heating ensures permanent adhesion with temperature and pressure varied for different resins. Pattern of dots vary from 3 to 12 dots per cm.

- For light weight fabrics small dots at high concentration are created and large dots in low concentration are created for heavy interlinings to ensure satisfactory bond formation.

Paste coating

- The paste coating is done by creating a paste using fine resin powder blended with water and other agents. The smooth paste is used to print on the base cloth.

- Significant amount of heat ensures removal of water and dots combine into solid resin. It provides precise shaped dots and is used in shirt collar fusibles.

Other methods

- Preformed method is used where preformed net is laminated to a base cloth to form precise dot patterns which are used in top collar fusible. Extrusion laminating is another method where a continuous film of polyethylene is produced. However, it results in a very stiff product like shirt collar.

- Emulsion coating method is used by dipping base cloth into a bath of emulsion, squeezing out excess resin by rolling and drying in oven to produce double-sided coatings.

6.8 Means of fusing

6.8.1 Temperature

- Temperature should be high enough to change the dry thermoplastic resin into a partially molten state as each resin has a limited range to achieve correct level of flow. Adhesion is affected by very low temperature and poor flow and very high temperatures cause strike through and strike back effects.

6.8.2 Pressure

- The equipment used must provide enough pressure to provide intimate contact between interlining and outer cloth. The equipment should ensure correct transfer of heat to the glue line and correct penetration of resin.

- When the pressure is too low, it causes adhesion and too high pressure results in excessive penetration causing strike back and strike through.

6.8.3 Time

- The fusing equipment must give enough time to allow temperature and pressure to induce melting of the resin and penetration of the outer fabric to produce a satisfactory bond. More time results in strike back and strike through.

- If thick fabric is used, it takes several seconds for resin to reach the required temperature. If temperature is not high enough then any extra time given will not enable the resin to soften and flow.

6.9 Fusing equipment

Fusing machines can be divided into three categories:

- Specialized fusing presses
- Hand irons
- Steam presses

6.9.1 Specialized fusing presses

Flat bed fusing machine

Fusing presses differ based on the way they operate and it affects both quality and productivity. Flat bed fusing press fusing machine uses two principles: (a) vertical action and (b) scissor action as shown in Figure 6.11.

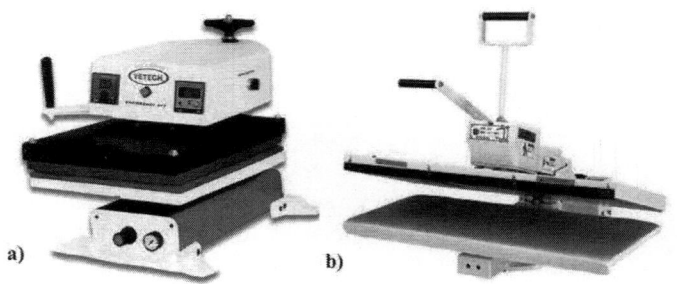

Figure 6.11   Flat bed Fusing machine types (a) Vertical action and (b) Scissor action.

The top plate is unpadded but the bottom plate has a resilient cover of silicone rubber, though it may be a felt pad. Both plates have an outer cover of PTFE, which can be cleaned easily to prevent straining and build-up of resin. Heat is provided by electric elements, usually in the top plate only, but sometimes in the bottom as well. The elements provide a uniform temperature over the whole surface.

It is aimed for a standard of control allowing a variation of 5°C either way from the required temperature over the whole surface area.

Pressure is applied by closing plates using following means:

(a) Mechanic

(b) Hydraulic

(c) Pneumatic

- Pressure system must be robust with accurate closing over large area. They must be free from distortion through heat, wear or mechanical faults.

- The vertical action has more accurate pressure than scissor action and fusing time is controlled by automatic timer with a normal cycle ranging between 8 s and 12 s.

- In the simplest mode of operation, the operator places the garment part face down on the lower plate, places the interlining resin side down on top of it in the correct position, and closes the press. This is slow and time consuming as the operator can do little that is productive for the duration of the fusing cycle.

- Variations:

    - Twin tray system with slides in and out from under top plate

    - Three section carousel with separate loading and unloading sections

- Advantages:

    - Small size

    - Relatively low cost

    - Reduces fabric shrinkage

- Disadvantages:

    - Slow and time consuming

    - Does not cover large area (not more than 1 m ′ 0.5 m)

    - Cannot be used for crush pile fabrics like velvet

Continuous fusing systems

In these types of fusing systems, garment parts with interlining are passed through a heat source and pressure is applied simultaneously as shown in Figure 6.12. Heat is provided in three ways:

Direct heating: Conveyor belt carries components into direct contact with heated surface. Drums or curved plates are employed.

Indirect heating: In this type of heating, components are carried through a heated chamber.

Low temperature:

- Components are carried through pre-heating zone

- Heating is either direct or indirect

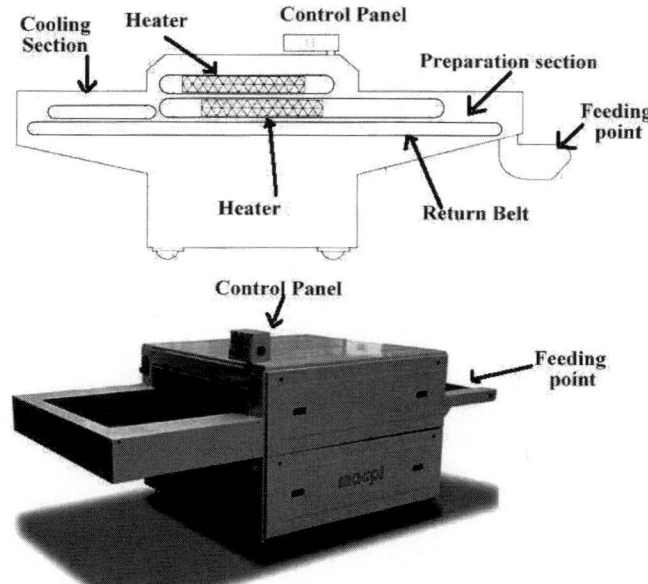

Figure 6.12   Continuous Fusing machine.

- Fusing takes place at 120°C

- It reduces heat shrinkage in outer fabric

- Temperature is well maintained in continuous fusing press

- On drum presses, tension of the conveyor belt presses components against the heated drum with nip rollers applying pressure while conveyor belts carry the components

- Pressure time is very small compared to flat bed presses

- The load to the rollers are applied by springs or pneumatically at ends

- Rubber covering of rollers are available in range of hardness, for example, shirt top collar fusing requires hardest rollers and outer wear fabrics require softer rollers

- Fusing time depends on speed of conveyor belts

Advantages:

- High quality

- High production

- Easy to operate

High frequency fusing

- In the fusing presses described so far, heat has been provided by electric heating elements. This limits the thicknesses of fabric that can be fused at once because of the time taken for the heat to transfer through the fabric to the resin. The heat may also produce shrinkage and colour changes. If multiple layers of fabric and interlining can be stacked up and fused simultaneously, productivity might be increased (Figure 6.13).

Figure 6.13   High frequency fusing machine.

- This method offers the possibility of eliminating shrinkage and colour changes. A high frequency generator provides alternating waves that are absorbed by certain types of polymer thereby generating heat.

- The plates of the press generate high frequency field and the heating effect is distributed uniformly in length and width and in the full height between the plates. This effect is known as dielectric heating.

- The heating effect is different for different polymers, and many fibres are not affected by them. The fusible adhesive material heats up much faster than either the interlining base fabric or the garment fabric.

- This results in bonding at the glue line without excessive heat being generated in the fabric. Multiple plies of garment and interlining can be stacked up to 70 mm high and since the heating effect is not dependent on its distance from a heat source, the adhesive at each joint should be raised to the same temperature.

- Less pressure is needed than with conventional fusing presses. The time required to generate the heat depends on the capacity of the high frequency unit and the weight of the load to be fused.

- The difficulty that arises with this method of fusing is that the press must be set to allow for the particular natural or man-made materials being used, the weight and thickness of those materials, and their moisture content.

- There should be due importance given to moisture content as wrong estimates may cause the press to over fuse and bond the whole stack together or under fuse and produce a poor bond on each garment part.

6.9.2 Hand iron

- The interlinings that can be fused at low temperature, low pressure and relatively short time use hand iron.

- It is a subjective method where the operator does not know the temperature at glue line and hence cannot apply uniform pressure (Figure 6.14).

Figure 6.14   Hand irons for fusing.

- It can iron only in small parts resulting in step-by-step heat transfer and is used in tailored jackets to fuse interlining sections temporarily.

- In jacket hems, slotted interlining tapes are fused and then ironed to hem along the line of slots.

6.9.3 Steam press

- It is the intermediate and final pressing of made up garments in which temperature at glue line achieved by steam.

The steam press depends on the steam pressure, press efficiency and cladding with pressure being applied mechanically or pneumatically. The lower part of press has vacuum for cooling (Figure 6.15).

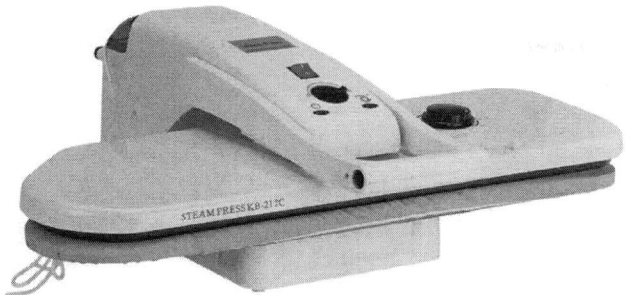

Figure 6.15 Steam Press for Fusing.

6.10 Methods of fusing

6.10.1 Single fusing

It is the safest and easiest method in which garment is placed on the right side facing down with the resin side on the garment.

6.10.2 Reverse fusing

In this type, the outer fabric lies on top of the fusible and is used in fusing shirt and blouse collars. In this type accurate positioning is difficult since interlining is smaller than the garment.

6.10.3 Sandwich fusing

It is effective on horizontal continuous press. The interlinings forming laminates are sandwiched between 2 outer fabrics (4 layers) which save time but preparation is longer and quality is unsatisfactory.

6.10.4 Double fusing

This is the fusing of two sorts of interlining to the outer fabric in one operation. It is most commonly used in shirt collars and men's jacket fronts. In shirt collar, one layer of fusible interlining skin is fused to the edge of the fabric, except on the neck edge and a second layer of fusible interlining patch is cut to fit within the sewing line. A 3 mm slot along the break line is fused on top of it. This double fusing process requires the achievement of two correct glue line temperatures in the two layers of resin.

6.11 Quality control in fusing

6.11.1 Temperature control

- Calibration is necessary before operating to relate actual glue line temperature to thermostat temperature.

- The glue line temperatures required for fusing will always be lower than the thermostat temperatures.

- Some extra amount of heat is required to remove the regain, the natural moisture contained within materials, which varies based on the situation. If machine is not fitted with sensor, following two methods can be used to check temperature. One by using portable pyrometer (a device with long wire inserted into press used on flat bed press) and the other, a thermo paper (a narrow strip with heat sensitive areas having a section marked with temperature and colour changes if temperature is reached while in press).

6.11.2 Pressure control

Pressure can be controlled by providing adequate and even pressure over the whole area of fusing press. A dial shows pressure (not actual pressure) between plates and only evenness can be measured. In continuous presses, strips of paper is passed through press and then stopped when partially through rollers. Then the strips are pulled by hand so that bowing in rollers can be checked.

6.11.3 Time control

The time on press controls must be related to actual time of fusing. The time is checked using a stop watch. Time cycle required to achieve desired temperature changes for different fabrics and linings. For example, thicker fabrics take more time for heat to penetrate into glue line.

Reference

1. Solinger, Jacob, *Apparel Manufacturing Handbook-Analysis, Principles and Practice*, Columbia Boblin Media Corp., 1988.

2. Carr, H. and Latham, B., *The Technology of Clothing Manufacture*, 2nd edn., Blackwell Scientific, Oxford, 1994.

3. Kunz, Grace I. and Glock, Ruth E., *Apparel Manufacturing: Sewn Product Analysis*, Prentice Hall, India, 2005.

4. Hayes, Steven, McLoughlin, John and Fairclough, Dorothy, *Cooklin's Garment Technology for Fashion Designers*, John Wiley and Sons, Ltd., Publication, UK, 2012.

5. http://www.macpi.it/.

6. http://www.macpiindia.com/.

7. http://www.vetech.in/.

8. http://www.hashima.co.jp/.

9. http://jinpu.en.ec21.com/.

10. http://capesewing.co.za/.

11. http://www.jesseheap.com/.

Packing machines

In this chapter, the different machines used in the apparel packing department are discussed. The machines used for the garment folding purpose, bagging, tagging, carton sealing and strapping machines are discussed along with the conveyors, which are generally used to transport the material in different forms at different places. Packing machines have become an integral part of apparel machinery as many machines have come up to reduce the human work and make packing an easy process. Such machines are discussed in this chapter.

Keywords: Package types, Garment folding machine, Bagging machine, Tagging machine, Vacuum packing

7.1   Packaging of products

- Packaging is an important part of the product, which has to receive a lot of attention to the people. Packaging is concerned with designing and producing of appropriate packages for a product.

- Packaging can be described as a coordinated system of preparing goods for transport, warehousing, logistics, sale and end use. Packaging means wrapping, compressing, filling or creating of goods for the purpose of protection of goods and their convenient handling.

Importance of packages

- **Physical protection** – Protection from mechanical shock, vibration, electrostatic discharge compression, temperature.

- **Barrier protection** – A barrier from oxygen water vapour, dust, etc.

- **Information transmission** – Packages and labels communicate how to use, transport, recycle or dispose the package or product.

- **Marketing** –Can be used by marketers to encourage potential buyers to purchase the product.

- **Security** – Packaging can play an important role in reducing the security risks of shipment.

- **Convenience** – Packages can have features that add convenience in distribution, handling, stacking, display, sale, opening, reclosing, use, dispensing, reuse, recycling and ease of disposal.

The main function of any package is

1. Distribution – Deals with packaging the apparel or allied product in a manner which permits the apparel product manufacturer to ship the product at lower cost and in shortest time to purchaser, without reducing the quality of the product.

2. Merchandising – Deals with presenting the apparel product in a manner designed to stimulate consumer desire for the product.

   - Both functions have same scope with respect to retaining the products durability and style specification during the journey from the manufacturer to retailer.

   - Creasing, crushing and dust are the quality deterrents that have to be prevented in packing practically in customer items.

7.2   Types of package

The basics types of package forms used in apparel and allied products are bags, boxes, cartons, cases, crates, twines and wrappers.

7.2.1 Merchandising package

- The product is packaged in the container in which the customer is expected to receive the product.

- It may be used to pack the product singly or multiple.

7.2.2 Shipping package

It is the package in which the **retailer receives** the product in bulk form.

- **Cases** – Made of wood, which has no opening in any of its sides (Figure 7.1)

- **Crates** – Made of wood, which has opening between the wood boards which make up its ends (Figure 7.2)

Figure 7.1   wooden cases.

Figure 7.2   Wooden crataes.

- **Box** – Made of cardboards/plastic – are container with separate covers. Box contains two or more pieces (Figure 7.3)

- **Cartons** – Made of cardboards/plastics – are containers which do not have separate covers. The carton is one piece container. Either folding type or set up (Figure 7.4)

Figure 7.3   Carton boxes made of plastics and card boards.

Figure 7.4   Carton boxes.

- **Bags** – Non-rigid container made from paper or plastic films. They do not have rigid structural form as like carton, cases, etc.

The two basic types are:

1. **Sacks** – Do not have flap –closed by folding all sides of opening (Figure 7.5)

Figure 7.5 Sacks – type of bag.

2. **Envelopes** – Non-rigid container with flap extension to close (Figure 7.6)

Figure 7.6 Envelope - a type of bag.

Wrap packages – These are parcels made by encasing the product in sheet paper or plastic film. The wrapping does not have the basic structure of a sack or envelope. Sacks and envelopes are usually fabricated before the packing. Wrappers are secured in the package formation with selling tape, cords or bands.

7.3 Packaging material

The basic packaging material used in the apparel and allied products are paper, plastic film, wood, nails staples, cords like twine and rope, gum tape (cloth and paper) and bands. Wood cases and crates used for bulky exports or shipments may be subjected to lot of shipment handling abuse. The wood consists of groove boards or sheet of plywood. Generally soft woods are used because they have sufficient strength and less cost than hardwoods.

Paper and films are the most popular packaging material in the apparel industries. The paper materials like kraft, crepe tissue, paper board, paper foil and water proof paper are used. Water proof papers are generally used for export packaging. Kraft paper is the most popular wrapping paper. It is natural brown paper made from sulphate pulp. Crepe papers are generally used for shock resistant packing. They are distinguished by their wrinkled surface. Fine crepe papers are also available for decorative packing of the merchandising.

Tissues are used for generally shock resistant and decoration purpose. Paper foils are metallic papers consisting of paper pulp laminated or coated with metals such as bronze, aluminium, etc. In apparel packing the foils are purely for decorative purpose. Plastic films are now replaced with all other packing materials than ever. Plastic films are made from cellulose derivatives, rubber bases or synthetic resins. They are ranging from transparent to opaque and also with respect to strength, it has superior strength than all other types of material.

7.4   Merchandising packaging

- The merchandise package is the unit the consumer receives when he/she selects the product. They are stimulating the sales of the product.

- From consumer point of view the merchandise package should:

  1. Identify the product

  2. Enhance the appeal of the product

  3. Attract the customer to the package

  4. Protect the product quality until the consumer uses the item

- **Transparent plastic film** is useful in meeting all of these requirements

- Viewing the product makes it easy to identify and attract the customer

- Colour and design on the package are the other ingredients that are used to identify, enhance and attract

- The manner in which the product is packaged geometrically is a big factor in enhancing the appeal of the garment

- The future utility of the package is the big factor in attracting the customer

- The artistic value of the package
- The ideal package should protect the product against sun, rain, wind, dust and moderate pressure

In retailer point of view

Merchandising package should have some additional values like

- Rise the consumer appeal
- Handle the cost of the package
- Managing space requirement of the package to store
- Identifying minimum time requirement for handling
  - For storage
  - Store selling area
  - Dispensing the package

A merchandise package must be designed to meet the needs of the retailer and the desires of the consumer. In many apparel items, merchandise package is reduced to a hanger and a transparent plastic film is used as a shoulder cover with a colourful tag on the garment. This is the ideal situation for many instances because the customer wants to feel the garment/fabric. If the consumer expects to check the drape or fit the garment, packing tags, pins or clamps should never impede the fitting process. A merchandise package that is ideal for one retailer is worthless for another retailer that is because their merchandising policies are different.

7.5 Shipment packing

The shipment package performs the distribution function. It is a package the carrier receives and delivers to the dealer or retailer. It delivers the merchandise packages to the retailer.

The criteria for a good shipping package are,

1. To protect and preserve garment quality
2. To reduce handling cost

In general the shipping packages are four types:

1. Closed container carrying goods – Cartons with hanger racks. This can be of two types as follows.

   (a) Covered completely individually by a merchant pack

(b)   Without a covering

2.   Open container carrying goods – Coats, suits or dresses transported with individual covers on hangers are the examples of open packages. It may be of both types as follows:

(a)   in closed merchandising packs

(b)   in open merchandising packs

However, any shipping package can carry open or closed merchandising package.

7.6   Packing equipment

7.6.1   Garment folding equipment

- They are used to fold the garments like shirt for enabling the packing of them. The focus of this equipment is to fold the garment and to reduce the cost of the garment folding operation along with improved production and packing quality. Figure 7.7 shows the high-speed automatic folding system designed to precisely fold a wide variety of garments and other textile items.

Figure 7.7 Continuous arment folding machine.

- The machine consists of folding plates, which electronically adjusts to the desired width while staying parallel to each other. This allows tighter, more consistent folding on all types of products.

- Wide transfer belts in the second fold area minimize maintenance and improve performance. The entire garment transport system uses a timing belt drive, which runs quieter and requires less long-term maintenance than chain and sprocket drives.

- Top covers are transparent and provide clear views and easy access to components from both sides of the folder. Quick release side panel covers allow easy access to the inner workings of the machine. However, the machine stops automatically if the side panels are open.

Special features of garment folding machine

- Fold plate sizes can be adjusted for different sizes
- Automatic sensors available to detect the garment, sensor increases production by automatically initiating the folding cycle when a sensor determines that a garment is in position on the drape table
- Electronic counter to monitor the production
- Short and long sleeve folding attachments

7.6.2 Card inserting machine

The automatic card feed/insert system is designed to integrate seamlessly with any automatic folding systems. These exceptionally fast card feeders are able to keep pace with the highest folder cycle speeds. They accept a wide variety of 16–20 point non-curled cardboard and chipboard stiffeners, which are fed into garments during the folding process. The card magazine can be refilled quickly to help maintain high productivity levels.

The top is also hinged so that the adjustable magazine can be quickly filled with 30 cm (12 in.) high stacks of cards. It replaces drape table and requires no additional floor space.

7.6.3 Garment stacking machine

- Stackers are the machines which are used to arrange the folded garments into stacks before the bagging and sealing process starts.

- Stacking machines can be fed directly from most automatic folders. They require minimal floor space and accept a wide variety of products up to 30 cm–38 cm in size, indexing user-set counts into uniform stacks up to 30 cm (12 in.) high as shown in Figure 7.8.

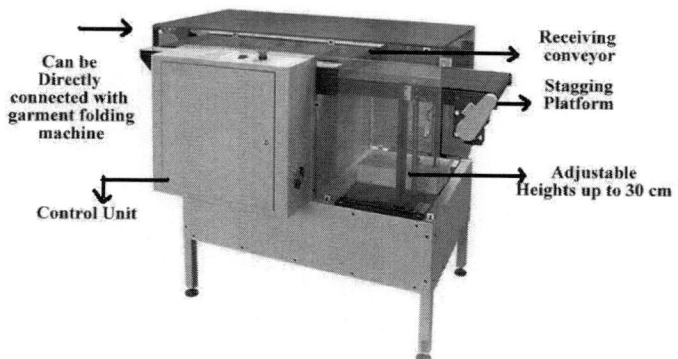

Figure 7.8   Garment stacking machine.

- Those stacks can then be sent to next machine in sequence (automatic bagging machine or onto conveyors). These machines stacking heights can be modified and enables the elevated discharge which frees operators from repetitive lowering and bending. The machine also comes with product thickness adjustment, variable feed speed, adjustable in-feed and out-feed heights and a programmable/mechanical counter.

Special feature

- Creates stacks of items up to 30 cm high and transports them to the next operation.

- Combines with bagging systems to produce multi-packs.

7.6.4 Garment loading/bagging equipment

- It loads the pre-fabricated container.

- The most economical loading machine in today's apparel industry is the loading of folded garment into the pre fabricated poly bags.

- In today's context, all the shaping and loading machines that are used for garment packing using plastic film are semi-automatic in nature. Hence, the speed and efficiency is based on the packer who handles the machine.

- Usage of fully automatic machine will increase the production rate as it will be totally based on the machine capacity.

The machine consists of two sections for bag loading and sealing system as shown in Figure 7.9.

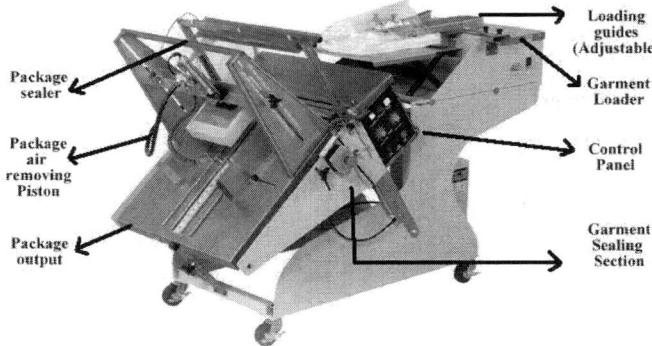

Figure 7.9   Garment packaging machine – Bagging and sealing of individual
garment.

Bag loading section

- The machine is capable of handling bags of various sizes.

- During the working phase, the plastic bag is opened with a blower by blowing the air, it enables the machine to create an efficient, ergonomic loading and sealing station.

- The bag loading section features an adjustable-width drop-load wicket holder and slide-and-lock bag table positioned to accommodate varied bag lengths.

- Pivoting entering guides (horns) are engineered to the product's shape to promote rapid bag loading.

- After loading and prior to sealing the bag, a pneumatic compression plate gently squeezes the bagged product to eliminate excess air and gives the product an attractive appearance.

The bag-sealing section

- Consists of adjustable seal-centering deck that can be quickly set to accommodate different product dimensions.

- The durable, high-temperature sealing element is temperature adjustable, and incorporates an automatic trim plate with an air pulse to push waste into a scrap bag.

- The seal bar holds the bag against the sealing element to create an attractive, airtight seal. If the seal bar encounters an obstruction on its way to the seal element, the seal bar will retract and the machine will automatically shut down and remain offline until manually reset.

Special features

- Any size of the garments can be packed

- Machine is very compact and easily movable

- Adjustable seal bar temperature and dwell time ensures fast and clean sealing of product

7.6.5 Garment transport conveyor

- The conveyor is generally used for transporting the products from one workstation to other workstation. The conveyors are in different types based on how they work. The garment conveyors generally have belt systems for transporting materials (Figure 7.10).

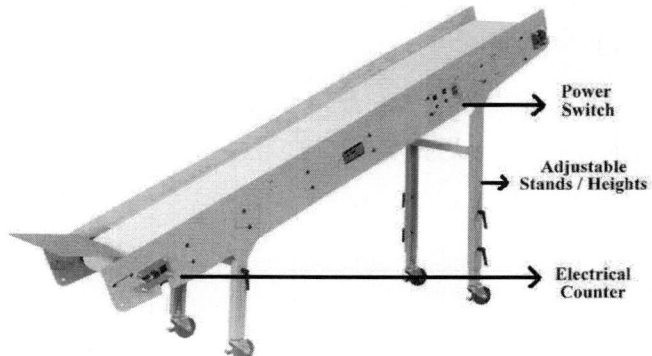

Figure 7.10   Garment transport conveyor.

- The freestanding incline conveyor will accept materials fed directly from any packaging machine.

- The delivery height and angle of the conveyor is adjustable. This allows the machine to be used in numerous situations requiring automated product conveyance, including those where products must be elevated so they can be dropped into boxes.

- The conveyor is available in 30 cm (12 in.) and 46 cm (18 in.) belt widths. A flat, rough-top belt provides reliable product transfer and allows pairing of the conveyor with automatic labeling machines.

- Based on the manufacturer, optional counter (either mechanical or electrical) can be fixed in the machine to measure the production, which can be reset by the operator.

- The conveyor can be set to stop once the preset number of count is reached.

7.6.6 Garment tagging machine

Tagging machines dramatically increase post-press productivity by attaching up to three product tags of varying sizes to a wide variety of items, including T-shirts, caps, socks, towels, sweatshirts and sweatpants. Since tag sets feed automatically, the machine is ready when the operator loads the next item, allowing the operator to concentrate on positioning the item on the machine and then stacking it after tagging.

- The machine can be operated by any operator irrelevant to the skill level

- For example, a price tag, a product information tag and a cleaning instruction tag can be automatically fed in any order onto the needle before being attached to the garment with a fastener.

- When only one tag is required, the machines allow the operator to stock each tag feeder with a different tag, making it possible to quickly change from one item to another by toggling the appropriate tag feeder (Figure 7.11).

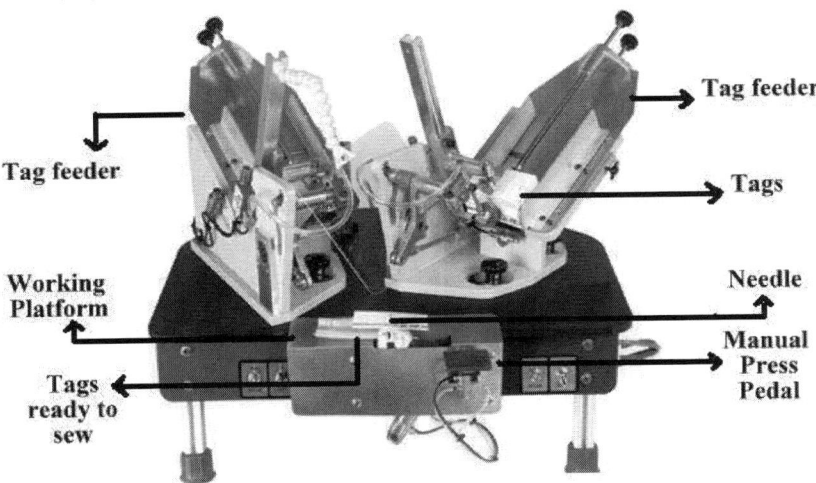

Figure 7.11 Garment tag attaching machine.

- When only one tag is required, the operator can stock each feeder with the same tag. The speed, efficiency and flexibility of taggers can produce a dramatic increase in operator productivity.

7.6.7 Container sealer

The container/carton sealing process in the apparel is either manual or using automated machines. In manual process the following instruments are used popularly. The sealing process can be done without any of these machines also.

7.6.7.1 Manual container sealer

1   **Stapler** – Used to seal the sides of carton

2.   Economical **sealing tape** dispenser and **banding machine**

   • Both used for sealing or securing cartons, as well as for the wrapped packages

   • Banding machines are sometimes used to reinforce the cases and crates

3.   **Plastic sealer** – Heat or electronic dispenser used to seal the end of plastic film bag. Container loader which is used to pack the garment in plastic film itself has plastic sealer as a part of it. The manual sealing equipment were given in Figure 7.12

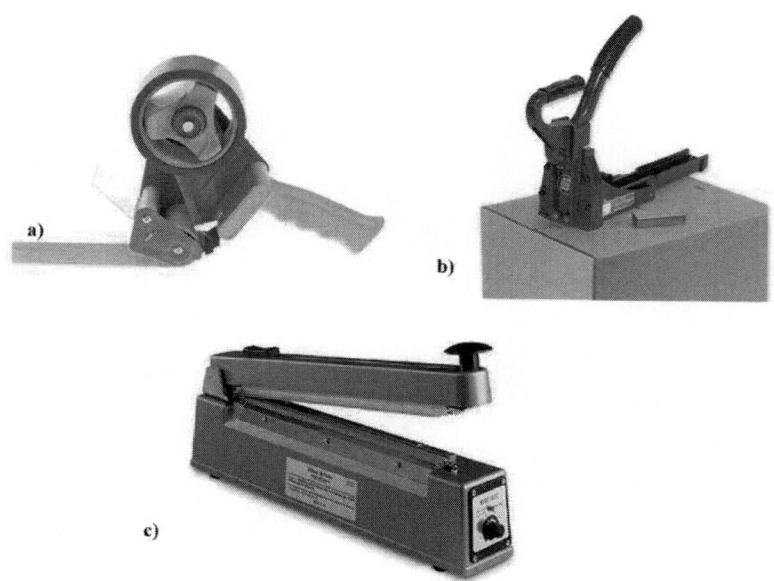

Figure 7.12   Manual packing machines (a) Tape dispenser (b) Stapler for carton box stapling and (c) Polyethylene bag sealing machine.

7.6.7.2 Automatic machine

The automatic carton box/container taping machines are generally used in the industries which handle large number of boxes per day in situations where the manpower is not affordable for the kind of jobs.

Automatic carton sealer is used with auto adhesive tape for the top and bottom sealing of American-type boxes for ensuring uniform carton size batches and enables manual dimensioning as shown in Figure 7.13.

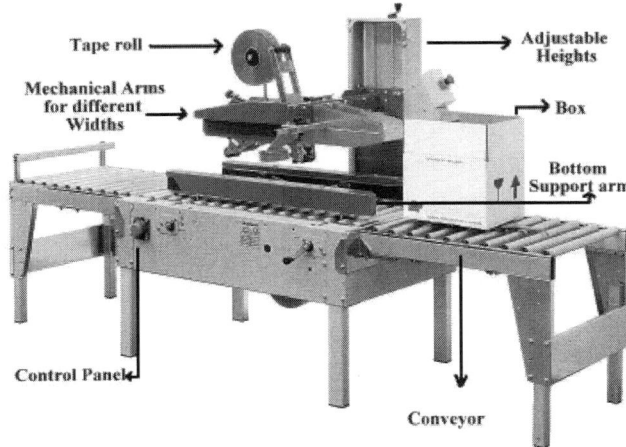

Figure 7.13 Automatic carton box sealing machine.

The adjustments are fast and easy, also enabling its use by non-specialist operators. The machine, if combined with in-feed and out-feed conveyors, represents a simple, fast and economical packaging station.

- Boxes are fed by means of 2 bottom side drive belts driven by a single motor. The regulations for the various formats are carried out by means of hand-wheels

- Fast taping of all types/sizes of cartons/boxes is possible and height is adjustable

- Easy operation, box just needs to be pushed and released

- Double motor (top and bottom) drive enables pushing the heaviest of boxes

- Production up to 1000 standard size boxes per hour is possible

- Average box transfer speed is 22 m/min

7.6.8 Carton/container strapping machine

Carton box strapping machines (Figure 7.14) are used in the case of apparel packing after the tape sealing of carton box in instances where the carton boxes were sealed with plastic or metallic strap wires to withstand the various transportation conditions.

- The strapping tension strength varies from 15 to 50 kg
- Adjustable strap width varies from 6 to 15 mm
- Strapping speed is 2.7 s/strap
- Package size can be of minimum 60 mm and maximum of any size
- Heat sealing method is used for the straps

Figure 7.14   Automatic Carton box strapping machine.

7.6.9 Container conveyor

There are various types of conveyors available for material handling purpose. These are the few types of equipment which are generally used in apparel industry to transport the container with in or out of shipping area. The basic systems used are:

(i)   Gravity chutes

A gravity conveyor moves the load without utilizing motor power sources, usually down an incline or through a person pushing the load along a flat conveyor as shown in Figure 7.15. The chute conveyor is one of the least expensive methods of conveying material. It is the simplest example of gravity-operated conveyor. Chute conveyor is used to provide accumulation in shipping areas; a spiral chute can be used to convey items between floors with minimum amount of space required. While the chute conveyors are economical, the main limitation of chute conveyors is the lack of control over the items being conveyed. The packages may tend to shift and turn so that jams and blockages occur.

Figure 7.15   Gravity Chute conveyor.

(ii) Roller conveyor

Gravity roller conveyors have metal rollers that roll on a fixed shaft as the rollers are spaced to allow free movement (Figure 7.16). The conveyor has a slight incline as gravity is used to move the object along its length. Gravity rollers can accommodate numerous objects placed on the conveyor at one time. By placing rollers closer together, you increase the capacity as well as provide a smoother rolling package.

Figure 7.16   Roller conveyors.

Live roller conveyors also have metal rollers where an object can be placed and transported to a desired location. Unlike gravity rollers, live rollers are mechanically powered using a belt or line shaft drive system. Most common speed ranges from 40 to 100 ft/min. Industries that need to start and stop the conveyor for merging and sorting products utilize live rollers in their operations for complete control during product transportation.

(iii) Belt or band conveyor

Belt conveyors are usually built with two or more pulleys, with a constant movement of the conveyor belt rotating to move product as given in Figure 7.17. One or both of the pulleys are powered, moving both the belt and the product on the belt onward. Belt conveyors can come with flat bed or on rollers. Each fills specific needs. Belt is almost always used to take product from one floor level to the next, either up or down. With lighter product, a limited use of rollers can also provide a quieter system.

Figure 7.17   Belt or band type conveyors.

7.6.10 Vacuum packing

The main functions of vacuum packaging are

- To reduce the shipping bulk of finished garment
- To reduce the shipping weight of the garments shipped
- To prevent a garment from accumulating dust or objectionable odours before and during shipping
- To prevent the garments from acquiring wrinkles or creases, during shipping which will have to be removed before the retailer displays the garment

- To minimize the storage space for both the manufacturer and retailer
- The vacuum package is not only used for the packing and storing garments. It is also used for packaging household accessories made from textiles, such as blankets, bedspreads, pillows and towels
- Any textile material with bulk structure can be easily compressed using vacuum package
- The moisture content is removed by passing the hung garment on the conveyor through the conditioning chambers which decrease the moisture content of the garment with hot, dry air, then cools it to a preset temperature in cool dry air
  - After that the garment is encased into a plastic film
  - This conditioning process permits the garment to be compressed with vacuum action without wrinkle or creases
  - When the vacuum pack is opened after it reaches the customer, it readily absorbs the necessary moisture to attain its normal stability
  - Thus it immediately sheds any wrinkles or creases formed during the vacuum packing process
  - The removal of moisture content prior to the vacuuming process reduces the wrinkle formation in packing. Because, in order to form a wrinkle or crease to have any retention, the fabric must have a certain degree of moisture
  - Vacuum packing is mainly used for tailored jackets. This pack prevents the creasing formation during the storage and transportation

7.6.11 Selecting the package design

Selection of the package is basically a sales function for both merchandising and shipment packaging. In general the package design selection is based on the following factors.

The package design depends on

- The sales policy of the firm
- The type of account the firm desires
- The type of merchant
- The nature of the product also limits the design
- The styling of items decides how it has to be packed
- The best way to show the garment to the customer

- The level of motivation required for the customer to ask for the product

- Some cases the customer may get attracted by hangers or open self display, etc.

7.7 Summary

In this chapter, the various packing machines used in the apparel industry are discussed. Clear illustrations are provided for easy understanding of the readers. The various merchandising packages available and importance of packaging is discussed in this chapter. The different packaging materials used in packing of garments are detailed. The packing equipments used are explained with sketches. Special features of the machines are described and the way to select the best package design is enlisted. This chapter will be very useful for readers to get a quick know how on the various machines used for packing.

References

1. Carr, Harold and Latham, Barbara, *The Technology of Clothing Manufacture*, 4th edition, Blackwell Publishing, Oxford, UK, 2000.

2. Solinger, Jacob, *Apparel Manufacturing Handbook-Analysis, Principles and Practice*, Columbia Boblin Media Corp., 1988.

3. http://www.mrprint.com/.

4. www.amscomatic.com.

5. http://www.watershed-packaging.co.uk/.

6. www.pactecindia.com.

7. www.new-delhi.all.biz.

8. www.flostor.com.

9. www.codingindia.com.